Springer Theses

Recognizing Outstanding Ph.D. Research

For further volumes:
http://www.springer.com/series/8790

Aims and Scope

The series "Springer Theses" brings together a selection of the very best Ph.D. theses from around the world and across the physical sciences. Nominated and endorsed by two recognized specialists, each published volume has been selected for its scientific excellence and the high impact of its contents for the pertinent field of research. For greater accessibility to non-specialists, the published versions include an extended introduction, as well as a foreword by the student's supervisor explaining the special relevance of the work for the field. As a whole, the series will provide a valuable resource both for newcomers to the research fields described, and for other scientists seeking detailed background information on special questions. Finally, it provides an accredited documentation of the valuable contributions made by today's younger generation of scientists.

Theses are accepted into the series by invited nomination only and must fulfill all of the following criteria

- They must be written in good English.
- The topic should fall within the confines of Chemistry, Physics, Earth Sciences, Engineering and related interdisciplinary fields such as Materials, Nanoscience, Chemical Engineering, Complex Systems and Biophysics.
- The work reported in the thesis must represent a significant scientific advance.
- If the thesis includes previously published material, permission to reproduce this must be gained from the respective copyright holder.
- They must have been examined and passed during the 12 months prior to nomination.
- Each thesis should include a foreword by the supervisor outlining the significance of its content.
- The theses should have a clearly defined structure including an introduction accessible to scientists not expert in that particular field.

Guglielmo Paoletti

Deterministic Abelian Sandpile Models and Patterns

Doctoral Thesis accepted by
the University of Pisa, Italy

 Springer

Author
Dr. Guglielmo Paoletti
LIP6—(Université Paris 6) UPMC
Paris
France

Supervisor
Prof. Sergio Caracciolo
Dipartimento di Fisica dell'Università
  di Milano
Milan
Italy

ISSN 2190-5053   ISSN 2190-5061   (electronic)
ISBN 978-3-319-34745-5   ISBN 978-3-319-01204-9   (eBook)
DOI 10.1007/978-3-319-01204-9
Springer Cham Heidelberg New York Dordrecht London

*a Raffaele
mio papá*

Supervisor's Foreword

The Abelian Sandpile Model is a particular cellular automaton: discrete state space and dynamical evolution on discrete time. A collection of grains of sand is distributed among the vertices of a graph. A vertex is unstable if it has at least as many grains as its degree. An unstable vertex topples by giving one grain to each neighbor vertex. So that also some neighbor vertex which was stable can become unstable. But the stable configuration which is reached is independent from the particular order of the topplings. The configuration obtained after relaxation is the same if you add some grain of sand first here and afterward there or in the reversed order. For this reason the Sandpile is called Abelian.

It was originally introduced on a finite portion of the square lattice in such a way that sand is added randomly and eventually deposited outside such a region, so that a stable configuration is always reached. By looking at the statistical properties of the recurrent configurations reached in this way, the model appears as the simplest case of self-organized-criticality, that is a dynamical system which shows algebraic long-range correlations, without any tuning of parameters.

The model has been soon generalized to a generic graph. And a remarkable correspondence between recurrent configurations and trees on the graph has been discovered.

Very complex and amazingly beautiful periodic patterns are often generated by the dynamics, in particular in deterministic protocols in which the sand is added in chosen sites.

In his work Paoletti derived exact results which are potentially interesting also outside the area of critical phenomena. Exact means also site by site and not on ensemble averages or coarse graining.

For example, Paoletti studies the appearance of allometric structures, that is patterns which grow in the same way in their whole body, and not only near their boundaries, as commonly occurs.

On the plane, beyond patches, that is patterns periodic in both dimensions, one dimensional periodic structures, called strings, show up. They are classified by their principal periodic vector k, that we call momentum. A simple relation between the momentum of a string and its density of particles, E, is obtained. This is reminiscent of a dispersion relation, $E = k^2$. Strings interact: they can merge and split and within these processes momentum is conserved. The modular group $SL(2, \mathbb{Z})$ plays an important role behind these laws. And this reveals interesting connections with number theory.

Indeed, this kind of analysis has already become interesting not only in out-of-equilibrium statistical mechanics, but also in mathematics, both in probability as in combinatorics.

New questions arise in an old subject. Therefore I believe that on one side this thesis can amuse computer practitioners for the simplicity in which charming patterns can be obtained, on the other it can attract the interest of researchers working in many different areas.

I wish to add a personal note. It has been a pleasure to have a student like Guglielmo and to work in collaboration with Andrea Sportiello on this beautiful subject.

Milan, April 2013 Prof. Sergio Caracciolo

Acknowledgments

Here we are, at the end of my Ph.D. Thesis but also at the end of four years of my life here in Pisa, the end of an *"era"*. It is funny how, simply changing the perspective, this is just the beginning of a new, big, adventure.

Acknowledgments are the only part of a Thesis every friend looks for when he open it. So, my friends, I am writing them for you and I hope you will like them, as much as I loved writing them.

First I want to thank my supervisor, Prof. Sergio Caracciolo, and my collaborator, Andrea S., who followed me wisely during these last few years. Their advises were fundamental in my research, as well as every discussion and dialogue we had made me grow. I feel very lucky to have had the possibility to work with them.

Then my thoughts go to my family, Raffaele, Adriana, and Rebecca, they all were so comprehensive to me, and I am sure that has been a tough task sometimes. They listened to my complaints, they spurred me on to greater efforts, and without the consciousness of their presence and their confidence in my skills I doubt I would have come to this point.

A special thank comes to Chiara and she knows why, cause any word would not be enough, a huge kiss (naic to you) and "Did I say that I need you"

How not say a word to Filippo who endured my presence and my good character in our three and more years as flatmates, and our friends and guest in via dell'Aeroporto 45: Emilio, Manu, Ale, Alberto, Sofia, Eleonora, Giangi, Davide, and Annalisa. Diners in via dell'Aeroporto are not finished, and I will join you once more.

Then Andrea, he is my Ph.D. mate. After our Ghezzano experience during these years we had the possibility to know each other, to travel around with the other Ghezzano boys, to party together with the German's guest and, last but not the least, to go lunch at "Circolino" (here is another place that I cannot forget and that I will bring in my memories wherever I will go, because "It is tough"). The lunches with Andrea, Katia, prof. Tonelli, Raffa, Elena will be object of many tales in the future. Thanks to you all and to "Circolino" I hope I will have the possibility to go there again whenever I will come back to Pisa.

Now, speaking about Ghezzano boys, a huge hug to Matteo, Damiano, and Claire, we all bring Ghezzano with us because it is not a place but a state of mind.

And also thanks for the good times together to Francesca, Umbe, Micche, Serena, Ciccio, Jacopo, Fabiolino, Laura, Lorenzo, Elisa, Ela, and Filippo.

During these last years in Pisa I must thanks the Caving group *GSPi*, cause there I made many friends, through them I discovered the Alpi Apuane and I also started to climb a bit. So thanks to Pascal, Giovanni (I will never forget our half marathon experience), Paolo, Andrea, Evelin, Teresa, il Pleb, Martina, Lara, Sandra, Il Giuntoli, Giorgio (il) Grande, Marco, Federico, Elena, Simone, and il Mancini.

Here in Pisa I also had the possibility to share my passion for photography with a group of amazing guys, with whom we founded a photo club "vividicontrasti", thanks to Cice (our artist), Fabio R., Fabio A., Nello, Gianni, Emmerre, lo Sfrek, Marta, Andrea, Chiara, Il Bini, Sara, Annalisa, and Dario.

I cannot forget to cite "Numeroundici" and Marco, the owner; in this little fantastic restaurant we had the first exhibitions of our photo club, but Marco is not only the guy behind the counter, he is a *friend* and I will always remember our chats and the photo sessions for the pictures of its website.

Although we all were far away, the old Milano aula *G* group is always in my mind, we met for holidays, parties around Europe, good times together in Milano, and I know we will meet soon, Marco, Luca Roma, Gio Viola, Jo, Ricky, Il Sanghi, Luca T., Elisa, and the Zanini brothers.

The same holds for all the guys from Varallo, you are always with me, wherever I go, I have been so happy every time I came back home to spend great time with you and I loved when you were guest at my place, I hope you will come also in my next destination! Thanks Paolo, Giovanni, Efrem, Cacaito, Elisa, Lucia, Nadia, Gigi, Laura, Andrea, Anna, Carla, Luz, Max, Agne, Jula, and Chicco.

I want also to thank all the physicists I met here in the Department and around the world in schools and congresses, we will meet again for sure, Bjarke, Walter, Matteo, Dario, Ezequiel, Benedetta, Carlos, Elia, Philippe, Roberto, Leone, Maurizio, Claudio, Vincenzo, Elio, and many others.

I already know I have forgot someone, someone who deserves a place in these Acknowledgments, please forgive me, I have a hug for all of you.

Guglielmo Paoletti

Contents

Chapter 1
Introduction

In the last twenty years, after the first article by Bak, Tang and Wiesenfeld [1] on
Self-Organized criticality (SOC), a large amount of work has been done trying to
better understand different features of this class of models. The most studied among
them is the *Abelian Sandpile Model* (ASM), that was actually proposed in that seminal
paper [1] as first archetype of SOC. The attention has been focused mainly on the
comprehension of the critical properties, in particular the determination of the critical
exponents of the avalanches. During my PhD I worked on the Abelian Sandpile
Model using unconventional approaches and focusing on not-standard features of
the model, related with the pattern formation that can be seen in the evolution of
particularly chosen configurations under deterministic conditions, that happened to
catch the attention of the scientific community only in the last few years.

1.1 Shape Formation in Cellular Automata

Since the appearance of the masterpiece by D'Arcy Thompson [2], there have been
many attempts to understand the complexity and variety of shapes appearing in Nature
at macroscopic scales, in terms of the fundamental laws which govern the dynamics
at microscopic level. Because of the second law of thermodynamics, the necessary
self organization can emerge only in non equilibrium statistical mechanics.

In the context of a continuous evolution in a differential manifold, the definition of
a shape implies a boundary and thus a discontinuity. This explains why catastrophe
theory, the mathematical treatment of continuous actions producing a discontinuous
result, has been developed in strict connection to the problem of *Morphogenesis* [3].
More quantitative results, and modelisations in terms of microscopic dynamics, have
been obtained by the introduction of stochasticity, as for example in the diffusion-
limited aggregation [4–7], where self-similar patterns with fractal scaling dimension
emerge [8], which suggest a relation with scaling studies in non-equilibrium.

G. Paoletti, *Deterministic Abelian Sandpile Models and Patterns*,
Springer Theses, DOI: 10.1007/978-3-319-01204-9_1,
© Springer International Publishing Switzerland 2014

Cellular automata, that is, dynamical systems with discretized time, space and internal states, were originally introduced by Ulam and von Neumann in the 1940s, and then they have been commonly used as a simplified description of non-equilibrium phenomena like crystal growth, Navier-Stokes equations and transport processes [9]. They often exhibit intriguing patterns and, in this regular discrete setting, *shapes* refer to sharply bounded regions in which periodic patterns appear. Despite very simple local evolution rules, very complex structures can be generated, and a scale unrelated to the lattice discretization can be produced spontaneously by the system evolution. The well-known Conway's Game of Life, which can even emulate an universal Turing machine, is an example of this emerging complexity, but a detailed characterization of such structures is usually not easy (see [10, 11], also for a historical introduction on cellular automata).

Cellular automata are defined through a set of configurations and an intrinsic evolution law. Given an automaton it may be studied under different evolution dynamics. These dynamics may be roughly divided into two general classes: *stochastic* and *deterministic* ones. As emphasized above, it is the ingredient of stochasticity that, in parallel to what emerges in critical phenomena for non-equilibrium statistical mechanics, suggests the possibility of obtaining probability laws which are "scale-free", that is, in which correlations among different spatially-separated parts do not decrease exponentially, at a scale related to the lattice discretization, but instead have an algebraic decay.

In equilibrium statistical mechanics, we expect criticality only at a fine-tuned value of a thermodynamic parameter (e.g., the temperature), if we have spontaneous symmetry breaking of a discrete group, and criticality for a subset of the degrees of freedom, in the whole ordered phase, if we have a continuous group and Goldstone bosons. In non-equilibrium systems the theoretical picture is less clear. Apparently, the non-equilibrium ingredient has often the tendency of increasing the scale-free region of parameters, or even automatically set the system in a scale-free point, without tuning of parameters, possibly by making all the "mass" operators irrelevant under the flow of the Renormalization Group (while, for equilibrium systems having a "magnetic" order parameter m, the temperature T and external field h always correspond to relevant operators). Such a feature is called *Self-Organized Criticality*.

A further feature of lattice automata is the possibility of producing *allometry*, that is a growth uniform and constant in all the parts of a pattern as to keep the whole shape substantially unchanged. Such a feature requires some coordination and communication between different parts of the system, and is thus at variance with diffusion-limited aggregation and other models of growing objects studied in physics literature so far, e.g. the Eden model, KPZ deposition and invasion-percolation [12–14], which are mainly models of aggregation, where growth occurs by accretion on the surface of the object, and inner parts do not evolve significantly. The lattice automaton discussed in [15], where this feature is discussed and outlined very clearly through an explicit model realization, is indeed a variant of the *Abelian Sandpile Model*, the general class of models that we will investigate in this thesis.

Another distinguished property of automata is that allometry emerges from a *deterministic* dynamics of the automaton, and thus we do not have a theoretical explanation in terms of stochastic processes, and steady-state distributions solving the master equation associated to a Markov chain. The theoretical approach followed in [15] is to investigate, at a coarse-grained level, the properties of what we may call a *discrete-valued Heat Equation*, strictly related to the intrinsic evolution law of the automaton (and not to the dynamics).

While the ordinary massless Heat Equation, formally solved within the realm of linear algebra, is automatically scale-free, the discretization in target space, and its counterpart, which is allowing for a tolerance offset in the resulting vector, produces a non-linearity. It is the interplay of these two ingredients that ultimately allows to have non-trivial shapes (through the non-trivial effects of non-linearity), on extended regions (through the absence of scale caused by the linear Heat Equation at leading order).

This phenomenology—of introducing a "small" non-linearity, through discretization, in a classical real-valued scale-free equation, and analyzing the possible emergence of morphogenesis—has a possibility of being realized in other models besides the Abelian Sandpile. Note however a subtlety here: in order to have a well-posed problem, it is crucial to find a discretization that preserves the unicity of the solution. In the case of the ASM, this relies on a non-trivial property of the model, namely, the unicity of the relaxation process, and ultimately the "abelianity" of the sandpile, that allows to perform a single relaxation in place of a sequence of maps in a non-commutative monoid, as is the case for a generic automaton.

1.2 The Many Faces of the Abelian Sandpile Model

As anticipated in the first paragraph, in this thesis we will concentrate on a particularly simple cellular automaton, the *Abelian Sandpile Model* (ASM). This model, now existing since about 20 years, was first proposed by Bak, Tang and Wiesenfeld in [1]. It has shown to be ubiquitous, as a toy model for a variety of features in Physics, Mathematics and Computer Science, including but not restricted to the ones described in the section above.

As all cellular automata, it has an intrinsic evolution law, and may be studied under different dynamics. These dynamics may be roughly divided into *stochastic* and *deterministic* ones. Thus we address explicit studies of the ASM under stochastic or deterministic dynamics as the *Stochastic Abelian Sandpile* and the *Deterministic Abelian Sandpile* respectively. We will give particular importance to the deterministic dynamics in the sandpile, although some particular stochastic dynamics will be part of the discussion.

The Stochastic Abelian Sandpile has been the first variant studied in physics literature. It is in this framework that it was proposed in the work of Bak, Tang and Wiesenfeld as first example of a model showing SOC[1] in 1987 because it shows

---

[1] A good introductory reading on Self Organized Criticality is the book *How nature works* [16] by Bak, which underlines the ubiquitous of SOC in nature, social sciences, economics and many other fields.

scaling laws without any fine-tuning of an external control parameter. This article has triggered a real avalanche. Indeed a huge number of articles has been published on SOC, the BTW article alone collected 1924 citations.[2] With the aim of determine the critical exponents of the model a great work has been done collecting statistics of the model under stochastic evolution, see [17]. Just a few years later, thanks to the work of Dhar, the algebraic properties of the sandpile were pointed out in [18, 19], where he first elucidated the group structure underlying the model. This structure allows to determine the statistical distribution of the Recurrent configurations of the model, that happens to be the uniform distribution. In a subsequent work in collaboration with Ruelle et al. [20] Dhar dealt in deep with the algebraic aspects of the sandpile, introducing some useful methods to determine the Smith normal form of the group.

The model was then studied by Creutz [21, 22], he was the first "looking" at the configurations and thus noting the particular and interesting shapes emerging in the identity configuration. He observed in particular how this configuration seemed to display self-similarity and a fractal structure [23, 24].

Afterwards the height's correlation of the model has been widely studied in a number of papers [25–27]. A connection with Conformal Field Theory came for the first time when the equivalence of the ASM with the $q \to 0$ limit of the q-state Potts Model was established [28]; thus two-dimensional ASM corresponds to a conformal field theory with central charge $c = -2$. The equivalence between the ASM and the Potts model in the limit $q \to 0$ gives as a side result a Monte Carlo algorithm to generate random spanning trees. Connections between ASM and the underlying CFT theory have been further studied by Ruelle et al. in a number of works [29–33] and then more recently in [34–37].

The connection with uniform spanning trees and the Kirchhoff theorem explains *a posteriori* the arising of self-organized criticality, i.e. the appearance of long-range behavior with no need of tuning any parameter. Indeed, uniform spanning trees on regular 2-dimensional lattices are a $c = -2$ logarithmic Conformal Field Theory (CFT) in [30], and have no parameter at all, being a peculiar limit $q \to 0$ of the Potts model in Fortuin-Kasteleyn formulation [38], or a limit of zero curvature in the OSP(1|2) non-linear σ-model [39]. If instead one considers the larger ensemble of spanning forests, in a parameter t counting the components (or describing the curvature of the OSP(1|2) supersphere), the theory in two dimensions is scale-invariant for three values: at $t = 0$ (the spanning trees, or the endpoint of the ferromagnetic critical line of Potts), at the infinite-temperature point $t = \infty$, and at some non-universal negative t corresponding to the endpoint of the anti-ferromagnetic critical line of Potts, being $t_c = -1/4$ on the square lattice. Through the correspondence with the non-linear σ-model, one can deduce at a perturbative level that the system is asymptotically free at $t = 0^+$, i.e. that the "coarse-graining" of a system with parameter $t > 0$, by a given scale factor Λ, "looks like" a system at parameter $t' > t$, but the functional dependence of $\ln\left(t'(t; \Lambda)/t\right)$ is only quadratic in t for $t \to 0$, instead of linear, as in the generic case [39, 40].

---

[2] Citations data at the 17th November 2011 as counted on the http://prl.aps.org.

The Deterministic Abelian Sandpile Model, especially at earlier times, has been studied in Mathematics and computer-science literature under the name of *Chip-Firing Game* [41, 42]. A particularly interesting result of chip-firing game in connection with Tutte polynomials has been obtained by Merino in [43], allowing to determine properties of the recurrent configurations of the model.

1.3 Overview

As anticipated in Sect. 1.1, in this thesis we want to study the ASM in connection with its capability to produce interesting patterns. This model is a surprising example of model that shows the emergence of patterns but maintains the property of being analytically treatable. Furthermore it is qualitatively different from other typical growth models—like Eden model, the diffusion limit aggregation, or the surface deposition [4, 13, 14]—indeed while in these models the growth of the patterns is confined on the surfaces and the inner structures, once formed, are frozen and do not evolve anymore, in the ASM the patterns formed grow in size but at the same time the internal structures acquire structure, as it has been noted in [15, 44–46].

There have been several earlier studies of the spatial patterns in sandpile models. The first of them was by Liu et al. [47]. The asymptotic shape of the boundaries of the patterns produced in centrally seeded sandpile model on different periodic backgrounds was discussed in [48]. Le-Borgne and Rossin [49] obtained bounds on the rate of growth of these boundaries, and later these bounds were improved by Fey-den Boer and Redig [50] and Levine and Peres [51]. An analysis of different periodic structures found in the patterns were first carried out by Ostojic [52] who also first noted the exact quadratic nature of the toppling function within a patch. Wilson [53] have developed a very efficient algorithm to generate patterns for a large numbers of particles added, which allows them to generate pictures of patterns with N up to 2^{26}.

There are other models, which are related to the Abelian Sandpile Model, e.g., the Internal Diffusion-Limited Aggregation (IDLA) [54], Eulerian walkers (also called the rotor-router model) [55–57], and the infinitely-divisible sandpile [51], which also show similar structure. For the IDLA, Gravner and Quastel showed that the asymptotic shape of the growth pattern is related to the classical Stefan problem in hydrodynamics, and determined the exact radius of the pattern with a single point source [58]. Levine and Peres have studied patterns with multiple sources in these models, and proved the existence of a limit shape [59]. Limiting shapes for the non-Abelian sandpile has recently been studied by Fey-den Boer and Liu [60].

The results of our investigation with intent of comprehend the patterns emerging in the ASM are reported along the thesis.

In Chap. 3 we will introduce some new algebraic operators, $a_i^\dagger$ and Π_i in addition to a_i, over the space of the sandpile configurations, that will become in the following basic ingredients in the creation of patterns in the sandpile. We derive some

Temperley-Lieb like relations they satisfy. At the end of the chapter we show how do they are closely related to multitopplings and which consequences has that relation on the action of Π_i on recurrent configurations.

In Chap. 4 we search for a closed formula to characterize the Identity configuration of the ASM. At this scope we study the ASM on the square lattice, in different geometries, and in a variant with directed edges, the *F-lattice* or *pseudo-Manhattan lattice*. Cylinders, through their extra symmetry, allow an easy characterization of the identity which is a homogeneous function. In the directed version, the *pseudo-Manhattan lattice*, we see a remarkable exact self-similar structure at different sizes, which results in the possibility to obtain a closed formula for the identity.

In Chap. 5 we reach the cardinal point of our study, here we present the theory of *strings* and *patches*. The regions of a configuration periodic in space, called *patches*, are the ingredients of pattern formation. In [15], a condition on the shape of patch interfaces has been established, and proven at a coarse-grained level. We discuss how this result is strengthened by avoiding the coarsening, and describe the emerging fine-level structures, including linear interfaces and rigid domain walls with a residual one-dimensional translational invariance. These structures, that we shall call *strings*, are macroscopically extended in their periodic direction, while showing thickness in a full range of scales between the microscopic lattice spacing and the macroscopic volume size. We first explore the relations among these objects and then we present full classification of them, which leads to the construction and explanation of a *Sierpiński* triangular structure, which displays patterns of all the possible patches.

References

1. P. Bak, C. Tang, K. Wiesenfeld, Self-organized criticality: an explanation of the 1/f noise. Phys. Rev. Lett. **59**, 381–384 (1987)
2. D. Thompson, *On Growth and Form*, 2nd edn. (Dover, New York, 1917)
3. R. Thom, *Structural Stability and Morphogenesis* (W.A. Benjamin, Reading, 1975)
4. T.A. Witten, L.M. Sander, Diffusion-limited aggregation, a kinetic critical phenomenon. Phys. Rev. Lett. **47**, 1400–1403 (1981)
5. T.A. Witten, L.M. Sander, Diffusion-limited aggregation. Phys. Rev. B **27**, 5686–5697 (1983)
6. P. Meakin, Diffusion-controlled cluster formation in 2-6-dimensional space. Phys. Rev. A **27**, 1495–1507 (1983)
7. T. Vicsek, *Fractal Growth Phenomena* (World Scientific, Singapore, 1989)
8. B. Mandelbrot, *The Fractal Geometry of Nature* (W. H. Freeman, New York, 1982)
9. B. Chopard, M. Droz, *Cellular Automata Modeling of Physical Systems* (Cambridge University Press, Cambridge, 1998)
10. S. Wolfram, *Cellular Automata and Complexity* (Addison-Wesley, Reading, 1994)
11. S. Wolfram, Statistical mechanics of cellular automata. Rev. Mod. Phys. **55**, 601–644 (1983)
12. M. Kardar, G. Parisi, Y.-C. Zhang, Dynamic scaling of growing interfaces. Phys. Rev. Lett. **56**, 889–892 (1986)
13. H. Herrmann, Geometrical cluster growth models and kinetic gelation. Phys. Rep. **136**, 153–224 (1986)
14. L. Barabasi, H. Stanley, *Fractal Concepts in Surface Growth* (Cambridge University Press, Cambridge, 1995)

15. D. Dhar, T. Sadhu, S. Chandra, Pattern formation in growing sandpiles. EPL (Europhys. Lett.) **85**, 48002 (2009)
16. P. Bak, *How Nature Works* (Copernicus, Secaucus, 1996)
17. S.S. Manna, Large-scale simulation of avalanche cluster distribution in sand pile model. J. Stat. Phys. **59**, 509–521 (1990). doi:10.1007/BF01015580
18. D. Dhar, Self-organized critical state of sandpile automaton models. Phys. Rev. Lett. **64**, 1613–1616 (1990)
19. D. Dhar, Sandpiles and self-organized criticality. Phys. A: Stat. Mech. Its Appl. **186**, 82–87 (1992)
20. D. Dhar, P. Ruelle, S. Sen, D.N. Verma, Algebraic aspects of abelian sandpile models. J. Phys. A: Math. Gen. **28**, 805 (1995)
21. M. Creutz, abelian sandpiles. Nucl. Phys. B (Proc. Suppl.) **20**, 758–761 (1991)
22. M. Creutz, abelian sandpiles. Comput. Phys. **5**, 198–203 (1991)
23. M. Creutz, P. Bak, Fractals and self-organized criticality, in *Fractals in Science*, ed. by A. Bunde, S. Havlin (Springer, Berlin, 1994), pp. 26–47
24. M. Creutz, Xtoys: cellular automata on xwindows. Nucl. Phys. B (Proc. Suppl.) **47**, pp. 846–849 (1996). http://thy.phy.bnl.gov/www/xtoys/xtoys.html
25. S.N. Majumdar, D. Dhar, Height correlations in the abelian sandpile model. J. Phys. A: Math. Gen. **24**, L357 (1991)
26. V.B. Priezzhev, Exact height probabilities in the abelian sandpile model. Phys. Scripta **1993**, 663 (1993)
27. V.B. Priezzhev, Structure of two-dimensional sandpile. I. Height probabilities. J. Stat. Phys. **74**, 955–979 (1994). doi:10.1007/BF02188212
28. S.N. Majumdar, D. Dhar, Equivalence between the abelian sandpile model and the $q \to 0$ limit of the Potts model. Phys. A: Stat. Mech. Its Appl. **185**, 129–145 (1992)
29. S. Mahieu, P. Ruelle, Scaling fields in the two-dimensional abelian sandpile model. Phys. Rev. E **64**, 066130 (2001), arXiv:0107150
30. P. Ruelle, A $c = -2$ boundary changing operator for the abelian sandpile model. Phys. Lett. B **539**, 172–177 (2002)
31. G. Piroux, P. Ruelle, Logarithmic scaling for height variables in the abelian sandpile model. Phys. Lett. B **607**, 188–196 (2005)
32. G. Piroux, P. Ruelle, Boundary height fields in the abelian sandpile model. J. Phys. A: Math. Gen. **38**, 1451 (2005)
33. G. Piroux, P. Ruelle, Pre-logarithmic and logarithmic fields in a sandpile model. J. Stat. Mech.: Theory Exp. **2004**, 10005 (2004)
34. M. Jeng, Conformal field theory correlations in the abelian sandpile model. Phys. Rev. E **71**, 016140 (2005)
35. S. Moghimi-Araghi, M. Rajabpour, S. Rouhani, abelian sandpile model: a conformal field theory point of view. Nucl. Phys. B **718**, 362–370 (2005)
36. M. Jeng, G. Piroux, P. Ruelle, Height variables in the abelian sandpile model: scaling fields and correlations. J. Stat. Mech.: Theory Exp. **2006**, P10015 (2006)
37. V. Poghosyan, S. Grigorev, V. Priezzhev, P. Ruelle, Pair correlations in sandpile model: a check of logarithmic conformal field theory. Phys. Lett. B **659**, 768–772 (2008)
38. A.D. Sokal, The multivariate Tutte polynomial (alias Potts model) for graphs and matroids, in *Surveys in Combinatorics*, ed. by B. Webb (Cambridge University Press, Cambridge, 2005)
39. S. Caracciolo, J.L. Jacobsen, H. Saleur, A.D. Sokal, A. Sportiello, Fermionic field theory for trees and forests. Phys. Rev. Lett. **93**, 080601 (2004)
40. S. Caracciolo, C. De Grandi, A. Sportiello, Renormalization flow for unrooted forests on a triangular lattice. Nucl. Phys. B **787**, 260–282 (2007)
41. A. Björner, L. Lovász, P. Shor, Chip-fring games on graphs. Eur. J. Combin. **12**, 283–291 (1991)
42. A. Björner, L. Lovász, Chip-firing games on directed graphs. J. Algebraic Comb. **1**, 305–328 (1992). doi:10.1023/A:1022467132614

43. C.M. Lopez, Chip firing and the Tutte polynomial. Ann. Comb. **1**, 253–259 (1997). doi:10.1007/BF02558479
44. T. Sadhu, D. Dhar, Pattern formation in growing sandpiles with multiple sources or sinks. J. Stat. Phys. **138**, 815–837 (2010). doi:10.1007/s10955-009-9901-3
45. D. Dhar, T. Sadhu, Pattern formation in fast-growing sandpiles, ArXiv e-prints (2011), arXiv:1109.2908v1
46. T. Sadhu, D. Dhar, The effect of noise on patterns formed by growing sandpiles, J. Stat. Mech.: Theory Exp. **2011**, P03001 (2011), arXiv:1012.4809
47. S.H. Liu, T. Kaplan, L.J. Gray, Geometry and dynamics of deterministic sand piles. Phys. Rev. A **42**, 3207–3212 (1990)
48. D. Dhar, The abelian sandpile and related models. Phys. A: Stat. Mech. Its Appl. **263**, 4–25 (1999). Proceedings of the 20th IUPAP International Conference on Statistical Physics
49. Y. Le-Borgne, D. Rossin, On the identity of the sandpile group. Discrete Math. **256**, 775–790 (2002). LaCIM 2000 Conference on Combinatorics, Computer Science and Applications.
50. A. Fey-den Boer, F. Redig, Limiting shapes for deterministic centrally seeded growth models. J. Stat. Phys. **130**, 579–597 (2008). doi:10.1007/s10955-007-9450-6
51. L. Levine, Y. Peres, Strong spherical asymptotics for rotor-router aggregation and the divisible sandpile. Potential Analysis **30**, 1–27 (2009). doi:10.1007/s11118-008-9104-6
52. S. Ostojic, Patterns formed by addition of grains to only one site of an abelian sandpile. Phys. A: Stat. Mech. Its Appl. **318**, 187–199 (2003)
53. D.B. Wilson, *Sandpile Aggregation Pictures on Various Lattices*. http://research.microsoft.com/en-us/um/people/dbwilson/sandpile/
54. G.F. Lawler, M. Bramson, D. Griffeath, Internal diffusion limited aggregation. Ann. Probab. **20**, 2117–2140 (1992)
55. V.B. Priezzhev, D. Dhar, A. Dhar, S. Krishnamurthy, Eulerian walkers as a model of self-organized criticality. Phys. Rev. Lett. **77**, 5079–5082 (1996)
56. Y.P. Lionel Levine, Spherical asymptotics for the rotor-router model in $\mathbb{Z}^d$. Indiana Univ. Math. J. **57**, 431–450 (2008)
57. A.E. Holroyd, L. Levine, K. Mészáros, Y. Peres, J. Propp, D.B. Wilson, Chip-firing and rotor-routing on directed graphs, in *In and Out of Equilibrium 2*, ed. by V. Sidoravicius, M.E. Vares, Progress in Probability, vol. 60 (Birkhäuser, Basel, 2008), pp. 331–364. doi:10.1007/978-3-7643-8786-0_17
58. J. Gravner, J. Quastel, Internal DLA and the Stefan problem. Ann. Probab. **28**, 1528–1562 (2000)
59. L. Levine, Y. Peres, Scaling limits for internal aggregation models with multiple sources. J. Anal. Math. **111**, 151–219 (2010). doi:10.1007/s11854-010-0015-2
60. A. Fey-den Boer, S.H. Liu, Limiting shapes for a non-abelian sandpile growth model and related cellular automata. Preprint (2010), arXiv:1006.4928v2

Chapter 2
The Abelian Sandpile Model

The State of the Art

It has been more than 20 years since Bak, Tang and Wiesenfeld's landmark papers on self-organized criticality (SOC) appeared [1]. The concept of self-organized criticality has been invoked to describe a large variety of different systems. We shall describe the model object of our interest: the *Abelian Sandpile Model* (ASM). The sandpile model was first proposed as a paradigm of SOC and it is certainly the simplest, and best understood, theoretical model of SOC: it is a non-equilibrium system, driven at a slow steady rate, with local threshold relaxation rules, which shows in the steady state relaxation events in bursts of a wide range of sizes, and long-range spatio-temporal correlations. The ASM consists of a special subclass of the sandpile models that exhibits, in the way we will discuss later, the mathematical structure of an abelian group, and its statistics is connected to that of spanning trees on the relative graph. There is a number of review articles on this subject, taking into account the connection of the model with the theme of SOC and its inner mathematical properties: Dhar [2], Priezzhev [3] and Redig [4, 5].

Here we present a review of the Sandpile Model theory based on the material than can be found therein with particular emphasis on mathematical aspects and on its stochastic dynamics; some further development given by us complete the review. This material will be necessary for the comprehension of the studies discussed in the following chapters.

2.1 General Properties

The ASM is defined as follows [6, 2]: we consider any (directed) graph $G = (V, E)$ with $|V| = N$ and vertices labeled by integers $i = 1, \ldots, N$, at each site we define a nonnegative integer height variable z_i, called the height of the sandpile, and a threshold value $\bar{z}_i \in \mathbb{N}^+$. We define an *allowed* configuration of the sandpile as a set $z \in \mathbb{N}^N$ of integer heights $z = \{z_i\}_{i \in V}$ such that $z_i \geq 0 \ \forall i \in V$; an allowed configuration $\{z_i\}$ is said to be *stable* if $z_i < \bar{z}_i \ \forall i \in V$. Therefore the set S of stable

G. Paoletti, *Deterministic Abelian Sandpile Models and Patterns*,
Springer Theses, DOI: 10.1007/978-3-319-01204-9_2,
© Springer International Publishing Switzerland 2014

configurations is $S = \bigotimes_{i \in V}\{0, \ldots, \bar{z}_i\}$. If we call $S_\pm \subset \mathbb{Z}^n$ the sets respectively such that $z_i \geq 0$ for all $i \in V$, and $z_i < \bar{z}_i$ for all $i \in V$. Then S^+ is the set of *allowed* configurations and is *stable* if it is in $S := S_+ \cap S_-$. The involution $z_i \to \bar{z}_i - z_i - 1$ exchanges S_+ and S_-.

The stochastic time evolution of the sandpile is defined in term of the *toppling matrix* Δ according to the following rules:

1. *Adding a particle*: Select one of the sites randomly, the probability that the site i is picked being some given value p_i, and add a grain of sand there. Obviously $\sum_i p_i = 1$. On addition of the grain at site i, z_i increases by 1, while the height at the other sites remains unchanged.
2. *Toppling*: If for any site i it happens that $z_i \geq \bar{z}_i$, then the site is said to be unstable, it *topples*, and lose some sand grains to other sites. This sand's grains transfer is defined in terms of an $N \times N$ integer valued toppling matrix Δ, which properties will be specified in (2.2a). On toppling at site i, the configuration z is updated globally according to the rule:

$$z_j \to z_j - \Delta_{ij} \quad \forall j \in V \tag{2.1}$$

If the toppling results in some other sites becoming unstable, they are also toppled simultaneously (it will be clear in the following that the order of toppling is unimportant). The process continues until all sites become stable[1] (we will see later under which conditions on the set of threshold values and Δ the final stability is guaranteed)

At each time step of the stochastic evolution, first we add a particle, as specified in rule 1, then we relax the configuration, that means to perform the necessary topplings to reach a stable configuration as stated in rule 2.

The *toppling matrix* Δ has the following properties:

$$\Delta_{ii} > 0, \ \forall i \in V \tag{2.2a}$$

$$\Delta_{ij} \leq 0, \ \forall i \neq j \tag{2.2b}$$

$$b_i^- := \sum_{j \in V} \Delta_{ij} \geq 0, \ \forall i \in V \tag{2.2c}$$

For future convenience we also define the integers

$$b_i^+ := \sum_{j \in V} \Delta_{ij}^T \geq 0 \tag{2.3}$$

for any vertex i.

We will adopt a *vector* notation for the collection of elements $\vec{\Delta}_i = \{\Delta_{ij}\}_{j=1,\ldots,N}$. With this notation it is possible to rewrite the *toppling rule* (2.1) for the toppling at site i as

---

[1] This process is called an *avalanche*

Fig. 2.1 A graphical representation of the general ASM. Each node denotes a site. On topplings at any site, one particle is transferred along each arrow directed outward from the site, each arrow corresponding to a unit in $-\Delta_{ij}$

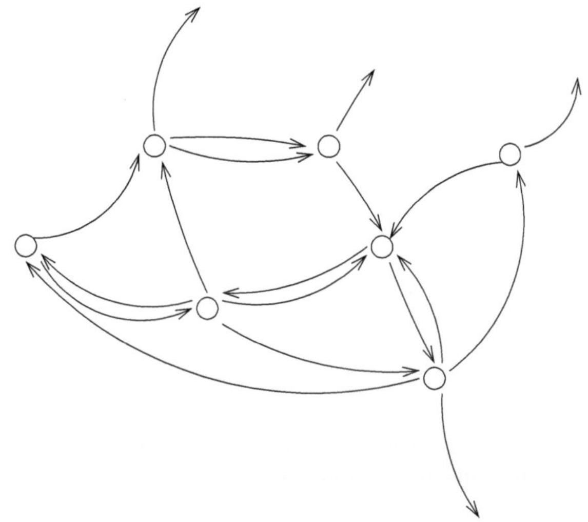

$$z \rightarrow z - \vec{\Delta}_i \tag{2.4}$$

The conditions (2.2a) just ensure that, on toppling at site i, z_i must decrease heights at other sites j can only increase and there is no creation of sand in the toppling process. In some sites could be possible to *lose* some sand during a toppling. If the graph G is undirected, the toppling matrix Δ is symmetric and $b_i^- = b_i^+ = b_i$.

The graph, in general directed and with multiple edges, is thus identified by the non-diagonal part of $-\Delta$, seen as an adjacency matrix, while the non-zero values b_i^- are regarded as *connections to the border*, and the sites i with nonzero b_i^- are said to be *on the border*, see Fig. (2.1). b_i^- is the total sand lost in the toppling process on i, so that, pictorially, we can think of this *lost sand* as dropping out of some boundary. b_i^+ is the difference between the amount of sand that can leave the site in a toppling and the amount that is added if all other sites would make a toppling. In the formulation on an arbitrary graph, as presented here, this concept of boundary does not need to correspond to any geometrical structure. We note that no stationary state of the sandpile is possible unless the particles can leave the system.

As an example, the original BTW model [1] is defined on an undirected graph which is a rectangular domain of the $\mathbb{Z}^2$ lattice. We have in this case

$$\Delta_{ij} = \begin{cases} +4 & \text{if } i = j \\ -1 & \text{if } i, j \text{ are nearest-neighbors} \\ 0 & \text{otherwise} \end{cases} \tag{2.5}$$

Here the connections with the border $b_i^+ = b_i^- = b_i$ are given, along the sides of the rectangle, by $b_i = 1$, while on the corners by $b_i = 2$. In this framework to be on the boundary (or in a corner) has a direct correspondence with the geometrical structure of the lattice.

We assume, without loss of generality, that $\bar{z}_i = \Delta_{ii}$ (this amounts to a particular choice of the origin of the z_i variables). Then we know that if a site i is stable, and the initial conditions for the heights are $z_i(t = 0) \geq 0 \ \forall i \in V$, at all times the allowed values for z_i are the ones for which holds $0 \leq z_i < \bar{z}_i$.

The procedure rule 1 followed by rule 2 defines a Markov chain on the space of stable configurations, with a given equilibrium measure. Running the stochastic dynamics for long times, that means after a large amount of sand added, the system reaches the stationary state.

As stated in rule 2, if a configuration is *unstable*, there is at least a vertex i where the configuration z has $z_i \geq \bar{z}_i$. The vertex i *topples* and the configuration z is updated following the rule (2.4). The new configuration reached after a toppling at site i is $t_i z = z - \vec{\Delta}_i$, where we call t_i the toppling operator at site i.

The collection of topplings needed to produce a stable configuration is called an *avalanche*. We shall assume that an avalanche always stops after a finite number of steps, which is to say that the diffusion is strictly *dissipative*. The size of avalanches can be studied statistically for interesting graphs (e.g. for a partition of $\mathbb{Z}^2$). In many cases of interest it seems to have a power law tail, which is signal of existence of long-range correlations in the system, see [7].

We shall denote by $\mathcal{R}(z)$ the stable configuration obtained from the relaxation of the configuration z, so $\mathcal{R}(z) \in S$ and

$$z \in S \quad \Leftrightarrow \quad z = \mathcal{R}(z) \, . \tag{2.6}$$

Given two configurations z and w we introduce the configuration $z + w$ which has at each vertex i the height $z_i + w_i$. Call $e^{(i)}$ the configuration which has non-vanishing height only at the site i where it has height 1, that is $e_j^{(i)} = \delta_{ij}$. Of course each configuration z can be obtained by deposing z_i particles at the vertex i

$$z_i e^{(i)} = \underbrace{e^{(i)} + e^{(i)} + \cdots + e^{(i)}}_{z_i} \tag{2.7}$$

so that summing on every vertex i, this means

$$z = \sum_{i \in V} z_i e^{(i)} \, . \tag{2.8}$$

2.1.1 Abelian Structure

Let $\hat{a}_i$ be the operator which adds a particle at the vertex i

$$\hat{a}_i z := z + e^{(i)} \tag{2.9}$$

then if z is not stable at the vertex j,

$$t_j \hat{a}_i z = \hat{a}_i t_j z \tag{2.10}$$

is easily verified.

Let now a_i be the addition of a particle at the vertex i followed by a sequence of topplings which makes the configuration stable. The stable configuration

$$a_i z = \mathcal{R}(e^{(i)} + z) \tag{2.11}$$

is independent from the sequence of topplings, because topplings commute. Indeed, the final configuration of a sequence of topplings does not depend on the order of unstable vertices chosen for each intermediate toppling. For this reason the model is said to be *abelian sandpile model* (ASM). More precisely, if the configuration z is such that $z_i > \bar{z}_i$ and $z_j > \bar{z}_j$ then

$$t_i t_j z = t_j t_i z \tag{2.12}$$

can be easily verified. Let us consider an unstable configuration with two unstable sites α and β, toppling first the site α leaves β unstable thanks to (2.2b), and, after the toppling of β, we get a configuration in which $z \rightarrow z - (\vec{\Delta}_\alpha + \vec{\Delta}_\beta)$ this expression is clearly symmetrical under exchange of α and β. Thus we get the same final configuration irrespective of whether α or β is toppled first. By repeated use of this argument we see that, in an avalanche, the same final state is reached irrespective of the sequence chosen for the unstable sites to topple. Similar reasoning apply for toppling of a site α followed by addition of a sand grain in β, so this gives the same result of the reverse ordered operation.

It is clear now that, applying two operators a_i and a_j, the configurations $a_j a_i z$ and $a_i a_j z$ coincide

$$a_i a_j z = a_j a_i z = \mathcal{R}(e^{(i)} + e^{(j)} + z) \tag{2.13}$$

so that a_i and a_j do commute, or in other word

$$[a_i, a_j] = 0 \quad \forall i, j \in V \tag{2.14}$$

Note that, while this property seems very general, it is not shared with most of the other SOC models, even other sandpile models, for example when the toppling condition depends on the gradient, in this case the order of toppling would matter, being the toppling rule not local and dependent on the actual height's values of the whole configuration.

Given two configurations z and z' we define an abelian composition $z \oplus z'$ as the sum of the local height variables, followed by relaxation

$$z \oplus z' = \mathcal{R}(z + z') = \left(\prod_{i \in V} a_i^{z_i} \right) z' = \left(\prod_{i \in V} a_i^{z_i'} \right) z \qquad (2.15)$$

and thus, for a configuration z, we define multiplication by a positive integer $k \in \mathbb{N}$:

$$k z = \underbrace{z \oplus \cdots \oplus z}_{k} \qquad (2.16)$$

The operators a_i's have some interesting properties. For example, on a square lattice, when 4 grains are added at a given site, this is forced to topple once and a grain is added to each of his neighbors. Thus:

$$a_j^4 = a_{j_1} a_{j_2} a_{j_3} a_{j_4} \qquad (2.17)$$

where j_1, j_2, j_3, j_4 are the nearest-neighbors of j.

In the general case one has, instead of (2.17),

$$a_i^{\Delta_{ii}} = \prod_{j \neq i} a_j^{-\Delta_{ij}}. \qquad (2.18)$$

by definition, because if we add $\bar{z}_i$ particles at the vertex i, after its toppling these particles move on the nearest-neighbor vertices, and, since now on, the relaxation to the stable configuration will be identical. Using the abelian property, in any product of operators a_i, we can collect together occurrences of the same operator, and using the reduction rule (2.18), it is possible to reduce the power of a_i to be always less than Δ_{ii}. The a_i are therefore the generators of a finite abelian semi-group (in which the associative property follow from their definition) subject to the relation (2.18); these relations define completely the semi-group.

Let now consider the repeated action of some given operator a_i on some configuration z. Since the number of possible states is finite, the orbit of a_i must close on itself, at some stage, so that $a_i^{n+p} z = a_i^n z$ for some positive period p, and non negative integer n. The first configuration that occurs twice in the orbit is not necessary z, so that the orbit consists of a sequence of transient configurations, followed by a cycle. If this orbit does not exhaust all configurations, we can take a configuration outside this orbit and repeat the process. So the space of all configurations is broken up into disconnected parts, each one containing one limit cycle.

Under the action of a_i the transient configurations are unattainable once the system has reached one of the periodic configurations. In principle the recurrent configurations might still be reachable as a result of the action of some other operator, say a_j, but the abelian property implies that if z is a configuration part of one of the limit cycles of a_i, then so is $a_j z$, in fact $a_i^p z = z$ implies that $a_i^p a_j z = a_j a_i^p z = a_j z$. Thus the transient configurations with respect to an operator a_1 are also transient with respect to the other operators $a_{j_1}, a_{j_2}, \ldots$, and hence occur with zero probability in the steady state. The abelian property thus implies that a_i maps the cycles of a_i

Fig. 2.2 graphical representation of the combined action
of a_i and a_j

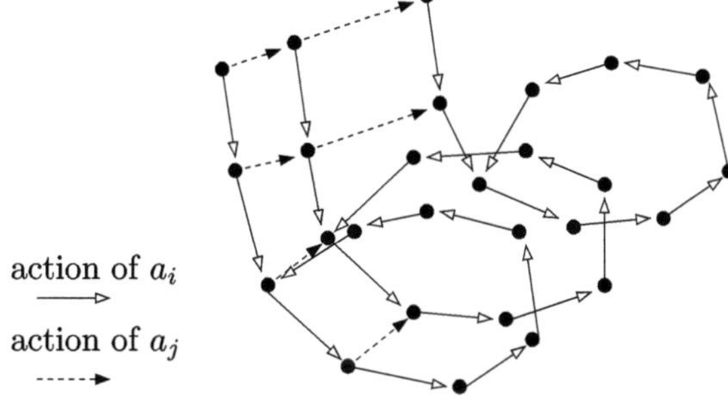

into cycles of a_i, and moreover that all this cycles have the same period Fig. (2.2). Repeating our previous argument we can show that the action of a_j on a cycle is finally closed on itself to yield a torus, possibly with some transient cycles, which may be also discarded. Continuing with the same arguments for the other cycles and other generators leads to the conclusion that the set of all the configurations in the various cycles form a set of multi-dimensional tori under the action of the a_i's.

The configurations that belong to a cycle are said to be *recurrent*, and can be defined, if we allow addition of sand with non zero probability in any site ($p_i > 0$ $\forall i$), as the configurations reachable by any other configuration with addition of sand followed by relaxation. We denote the set of all recurrent configurations as **R** and the set of the transient ones as **T**.

Given the natural partial ordering, $z \preccurlyeq z'$ iff $z_i \leq z_i'$ for all i, then **R** is in a sense "higher" than **T**, more precisely

$$\nexists\, (z, z') \in \mathbf{T} \times \mathbf{R} : \qquad\qquad z \succ z' ; \qquad\qquad (2.19)$$

$$\forall z \in \mathbf{T} \;\; \exists z' \in \mathbf{R} : \qquad\qquad z \prec z' . \qquad\qquad (2.20)$$

In particular, the *maximally-filled* configuration $z_{max} = \{\bar{z}_i - 1\}$ is in **R**, and higher than any other stable configuration.

2.2 The Abelian Group

The set **R** of recurrent configurations is special. Indeed in **R** it is possible to define the inverse operator a_i^{-1} for all i, as each configuration in a cycle has exactly one incoming arrow corresponding to the operator a_i. Thus the a_i operators generate a group. The action of the a_i's on the states correspond to translations of the torus. From the symmetry of the torus under translations, it is clear that all recurrent states occur in the steady state with the same probability.

This analysis, which is valid for every finite abelian group, leaves open the possibility that some recurrent configurations are not reachable from each other, in which case there would be some mutually disconnected tori. However, such a situation cannot happen if we allow addition of sand at all sites with non zero probabilities ($p_i > 0 \; \forall i$). Let us define z_{max} as the configuration in which all sites have their maximal height, $z_i = \Delta_{ii} - 1 \quad \forall i$. The configuration z_{max} is reachable from every other configuration, is therefore recurrent, and since inverses a_i's exist for configurations in **R**, every configuration is reachable from z_{max} implying that every configuration lie in the same torus.

Let $\mathcal{G}$ be the group generated by operators $\{a_i\}_{i=1,\dots,N}$. This is a finite group because the operators a_i's, due to (2.18), satisfy the closure relation:

$$\prod_{i=1}^{N} a_i^{\Delta_{ji}} = I \qquad \forall i = 1, \dots, N \tag{2.21}$$

the order of $\mathcal{G}$, denoted as $|\mathcal{G}|$, is equal to the number of recurrent configurations. This is a consequence of the fact that if z and z' are any two recurrent configurations, then there is an element $g \in \mathcal{G}$ such that $z' = gz$. We thus have:

$$|\mathcal{G}| = |\mathbf{R}| \tag{2.22}$$

Given the group structure, the semi-group operation (2.15) is raised to a group operation, in particular the composition of whatever z with the set **R** acts as a translation on this toroidal geometry. A further consequence is that, for any recurrent configuration z, the inverse configuration $-z$ is defined, and $k\,z$ is defined for $k \in \mathbb{Z}$ too.

The identity of this abelian group, denoted by Id_r, is called recurrent identity or *Creutz identity* after Creutz first studies in [8, 9] and is the only stable recurrent configuration such that

$$\forall z \in \mathbf{R} \quad Id_r \oplus z = z \tag{2.23}$$

2.3 The Evolution Operator and the Steady State

We consider a vector space $\mathcal{V}$ whose basis vectors are the different configurations of **R**. The state of the system at time t will be given by a vector

$$|P(t)\rangle = \sum_z \mathsf{Prob}(z, t) \, |z\rangle \, , \tag{2.24}$$

where $\mathsf{Prob}(z, t)$ is the probability that the system is in the configuration z at time t. The operators a_i can be defined to act on the vector space $\mathcal{V}$ through their operation on the basis vectors.

The time evolution is Markovian, and governed by the equation

$$|P(t+1)\rangle = \mathcal{W} |P(t)\rangle \tag{2.25}$$

where

$$\mathcal{W} = \sum_{i=1}^{N} p_i a_i \tag{2.26}$$

To solve the time evolution in general, we have to diagonalize the evolution operator $\mathcal{W}$. Being mutually commuting, the a_i may be simultaneously diagonalized, and this also diagonalizes $\mathcal{W}$. Let $|\{\phi\}\rangle$ be the simultaneous eigenvector of $\{a_i\}$, with eigenvalues $\{e^{i\phi_i}\}$, for $i = 1, \dots N$. Then

$$a_i |\{\phi\}\rangle = e^{i\phi_i} |\{\phi\}\rangle \quad \forall i = 1, \dots, N. \tag{2.27}$$

We recall that the a_i operators now satisfy the relation (2.21). Applying the l.h.s. of this relation to the eigenvector $|\{\phi\}\rangle$ gives $\exp(i \sum_j \Delta_{kj}\phi_j) = 1$, for every k, so that $\sum_j \Delta_{kj}\phi_j = 2\pi m_k$, or inverting,

$$\phi_j = 2\pi \sum_k \left[\Delta^{-1}\right]_{jk} m_k , \tag{2.28}$$

where Δ^{-1} is the inverse of Δ, and the m_k's are arbitrary integers.

The particular eigenstate $|\{0\}\rangle$ ($\phi_j = 0$ for all j) is invariant under the action of all the a's, $a_i |\{0\}\rangle = |\{0\}\rangle$. Thus $|\{0\}\rangle$ must be the stationary state of the system since

$$\sum_i p_i a_i |\{0\}\rangle = \sum_i p_i |\{0\}\rangle = |\{0\}\rangle . \tag{2.29}$$

We now see explicitly that the steady state is independent of the values of the p_i's and that *in the steady state all recurrent configurations occur with equal probability.*

2.4 Recurrent and Transient Configurations

Given a stable configuration of the sandpile, how can we distinguish between transient and recurrent configurations? A first observation is that there are some forbidden subconfiguration that can never be created by addiction of sand and relaxation, if not already present in the initial state. The simplest example on the square lattice case is a configuration with two adjacent sites of height 0, $\boxed{0\,|\,0}$. Since $z_i \geq 0$, a site of height 0 can only be created as a result of toppling at one of the two sites (toppling from anywhere else can only increase his height). But a toppling of either of this sites results in a height of at least 1 in the other. Thus any configuration which contains two adjacent 0's is transient. With the same argument it is easy to prove that the following configurations can never appear in a recurrent configuration:

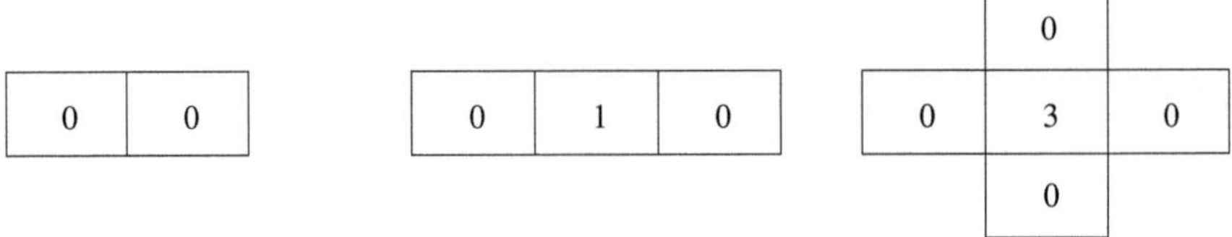

Fig. 2.3 Examples of forbidden subconfigurations

In general a *forbidden subconfiguration* (FSC) is a set F of r sites ($r \geq 1$), such that the height z_j of each site j in F is less of the number of neighbors than j in F, precisely:

$$z_i < \sum_{j \in F \setminus \{i\}} (-\Delta_{ij}) \quad \forall i \in F \tag{2.30}$$

The proof of this assertion is by induction on the number of sites in F. For example the creation of the $\boxed{0\;\;1\;\;0}$ subconfiguration must involve toppling at one of end sites, but then the subconfiguration must have had a $\boxed{0\;\;0}$ before the toppling, and this was shown before to be forbidden.

An interesting consequence of the existence of forbidden configurations is the following: consider an ASM on an undirected graph, with N_b bonds between sites, then in any recurrent configuration the number of sand grains is greater or equal to N_b. Here we do not count the boundary bonds, corresponding to particles leaving the system. To prove this, we note that if the inequality is not true for any configuration, it must have a FSC in it.

2.4.1 The Multiplication by Identity Test

Consider the product over sites $i \in V$ of equations (2.17)

$$\prod_i a_i^{\Delta_{ii}} = \prod_i \prod_{j \neq i} a_j^{-\Delta_{ij}} = \prod_i a_i^{-\sum_{j \neq i} \Delta_{ji}} \tag{2.31}$$

On the set $\mathbf{R}$, the inverses of the formal operators a_i are defined, so that we can simplify common factors in (2.31), recognize the expression for b_i^+, and get

$$\prod_i a_i^{b_i^+} = I \tag{2.32}$$

so that $\prod_i a_i^{b_i^+} z = z$ is a necessary condition for z to be recurrent (but it is also sufficient, as no transient configuration is found twice in the same realization of the

Markov chain), and goes under the name of *identity test*. If we denote by b^+ the configuration

$$b^+ = \sum_{i \in V} b_i^+ e^{(i)} \tag{2.33}$$

the identity test means that

$$z \in \mathbf{R} \Leftrightarrow z = z \oplus b^+ \tag{2.34}$$

In the next section we will see how to obtain the same result in an easier and faster way, without actually performing the addition and relaxation.

This relation gives information also on the recurrent identity itself. Indeed it says that if $b^+ \in \mathbf{R}$ then $b^+ = Id_r$. Otherwise there must be a positive integer k such that

$$\forall \ell \in \mathbb{N} : \ell \geq k \quad \underbrace{b^+ \oplus b^+ \oplus \cdots \oplus b^+}_{\ell} = Id_r \tag{2.35}$$

then

$$Id_r = kb^+. \tag{2.36}$$

2.4.2 Burning Test

We define a simple recursive procedure to discover if a configuration is recurrent, based on the mechanical check of presence or absence of FSC in the configuration. We consider at each step a test set, say T, of sites. At the beginning T consists of all the sites of the lattice we are considering; we test the hypothesis that T is a FSC using the inequalities (2.30). If these inequalities are satisfied for all sites in T, then the hypothesis is true, T has a FSC, and the configuration in exam is transient. Otherwise there are some sites for which the inequalities are violated, these sites cannot be part of any FSC, in fact the inequalities will remain unsatisfied even though T is replaced by a smaller subset of sites. We delete these sites from T and we have a new subset T', we say we *burn* these sites while the remaining are *unburnt*; at this point we repeat the procedure to check whether T' is a FSC. We follow this scheme until we cannot burn anymore site. If we are left with a finite subset F of unburnt sites this is a FSC and the configuration is transient, if the set of unburnt sites eventually becomes empty the configuration in exam is found to be recurrent. We call the procedure just presented the *burning test*.

In the *burning test* it does not matter in which order the sites are burnt. It is however useful to introduce the concept of time of burning and to add to the graph a site, named *sink*, which is connected to all the "boundary" sites with as many links as the number of lost particles in a toppling by the boundary sites in exam, it never topples and only collects sand. There is a natural way to choose a time of burning for each site. At time $t = 0$, all the sites in $T^{(0)} = T$ are unburnt except the sink. At any time t a site is called *burnable* iff the inequality (2.30) is unsatisfied with respect to the set $T^{(t)}$ reached at that time, then a burnable site at time t becomes burnt at time $t + 1$, and so remains for the successive times. With this prescription we label each site of the graph with a burning time, depending just on the configuration in exam.

We want now to draw a path for the "fire" to propagate to the whole graph, starting from the sink. Take an arbitrary site i, except the sink. Let $\tau_i + 1$ be the time step at which this is burnt, then the burning rule implies that at time τ_i at least one of his neighbor sites has been burnt. Let r_i be the number of such neighbors and let us write:

$$\xi_i = \sum_{j=1}^{r_i} (-\Delta_{ij}) \tag{2.37}$$

where the primed summation runs over all unburnt neighbors of i at time τ_i. Then we have $z_i \geq \xi_i$ since the site i is burnable at time τ_i; but, since it was not burnable at time $\tau_i - 1$ we must also have:

$$z_i < \xi_i + K \tag{2.38}$$

where K is the number of bonds linking i to his neighbors which were unburnt at time τ_i. During the burning test we say that fire reaches the site i by one of the K bonds. Obviously when $K = 1$ there is just one possibility so there are no problems, and we say that the fire reaches i from the only site possible. If $K > 1$ we have to select one bond through which the fire reaches i depending on his height z_i. For this purpose we order the bonds converging on the site i in some sequence (e.g. $\{(i, i_1), (i, i_2), (i, i_3), \ldots\}$), the order is arbitrary and can be chosen independently on each site i. Now we can write $z_i = \xi_i + s - 1$ for some $s > 0$, we say that fire reaches i using the s-th link in the ordered list of the possible ones. This procedure gives a unique path for the fire to reach each site i, given the configuration of heights in the sandpile and the prescription on the order of bonds converging on each site. The set of bonds along which fire propagates, connects the sink with each site in the graph, and there are no loops in each path. Thus the set just obtained is a spanning tree on the graph $G' = G + \{sink\}$.

Choosing a particular prescription for a given graph we can obtain for each recurrent configuration a unique spanning tree. For example on a square lattice we have four bonds for each site, let us call them N-E-S-W, where the cardinal points denotes the direction of incidence, we can choose the prescription N>E>S>W and obtain for a recurrent configuration the corresponding spanning tree. In Fig. (2.4) is shown an example of burning test for a 4×4 square lattice, with prescription NESW.

Let us note a few facts about the burning test before going on to the next section. Although all recurrent configurations were shown in [6] to pass the burning test; conversely, it was shown in [10] only for sandpiles with symmetric toppling rules –which is if the same amount of sand is transferred to site j when site i topples as it is transferred to i when j topples– that all stable configurations which pass the test are recurrent. The burning test is not therefore valid in general, indeed there exist simple asymmetric sandpiles having stable configurations which pass the test but are

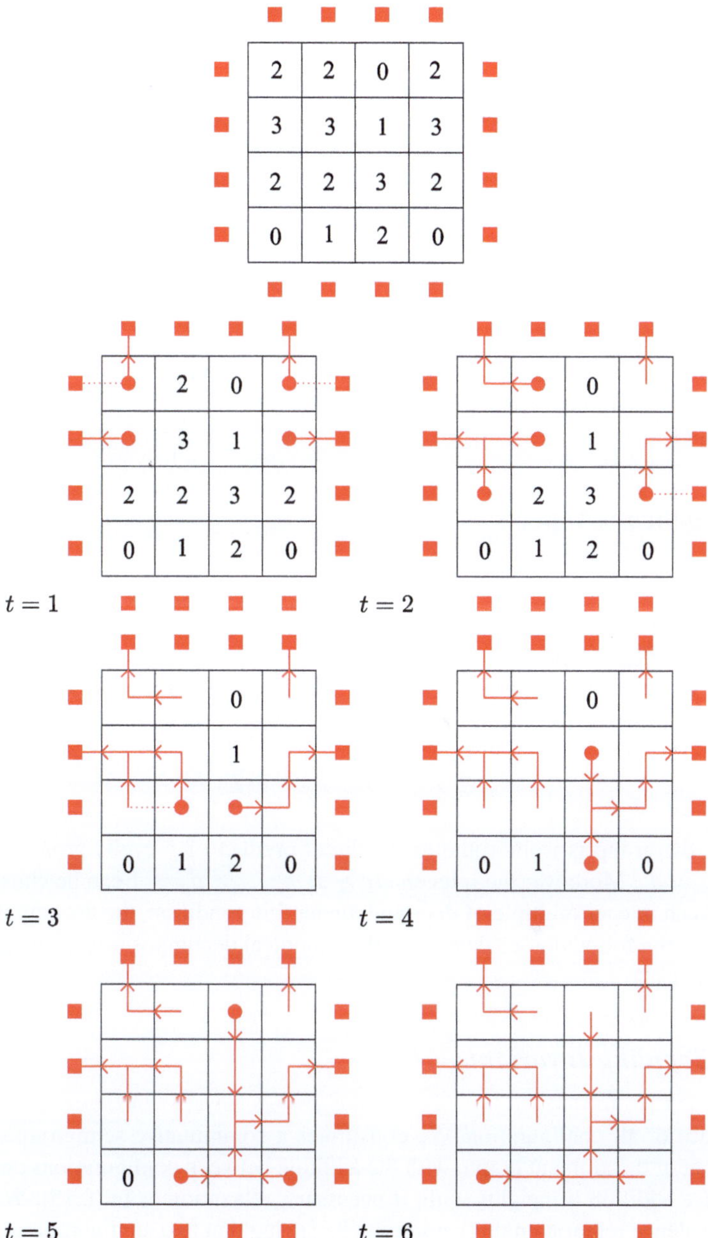

Legend: with ■ are denoted the sites such that, connected together, represent the sink, with ● the sites burnt at each step of the algorithm and with ········· , the bonds through which the fire could have reached the site but were rejected.

Fig. 2.4 Example of burning test acting on a given configuration. At each time is displayed the progress of the algorithm until at t=6 all sites become "burnt"

not recurrent; i.e. in the ASM with toppling matrix $\Delta = \begin{pmatrix} 4 & -3 \\ -1 & 2 \end{pmatrix}$, then there are $\det \Delta = 5$ recurrent configurations $(2, 0)$, $(3, 0)$, $(1, 1)$, $(2, 1)$ and $(3, 1)$ and satisfy the burning test, but $(1, 0)$ passes the burning test even if it is not recurrent. A site like 2 which has more incoming arrows than outgoing arrows is called *greedy* or *selfish*. In this case when adding a frame identity to the configuration, then some sites topple twice, and this makes the burning test fail as it assumes that under multiplication by the identity operator (2.32), each site topples only once.

This gap has been filled by Speer that in [11] introduced the *script test*, which is a generalization of the burning test valid in case of asymmetric sandpiles. Sandpile configurations which pass the script test are precisely the recurrent configurations of a sandpile. For sandpiles without greedy sites, the script test reduces to the burning test even for asymmetric sandpiles.

2.5 Algebraic Aspects

We report here some features of the abelian group $\mathcal{G}$ associated to the ASM. In particular we determine scalar function, invariant under toppling, and the rank of the group for the square lattice.

First we recall that any finite abelian group $\mathcal{G}$ can be expressed as a product of cyclic groups in the following form:

$$\mathcal{G} \cong \mathbf{Z}_{d_1} \times \mathbf{Z}_{d_2} \times \cdots \times \mathbf{Z}_{d_g} \tag{2.39}$$

That is, the group is isomorphic to the direct product of g cyclic groups of order $d_1, d_2, \ldots, d_g$. Moreover the integers $d_1 \geq d_2 \geq \ldots \geq d_g > 1$ can be chosen such that d_i is an integer multiple of d_{i+1} and, under this condition, the decomposition is unique. In the following we determine the canonical decomposition of the group.

2.5.1 Toppling Invariants

The space of all configurations S_+ constitutes a commutative semigroup over the vertex-set of the ambient graph, with the addition between configurations defined as a sitewise addition of heights with, if necessary, relaxation as in (2.15). We define an equivalence relation on this semigroup by saying that two configurations z and z' are equivalent iff there exists $N = |V|$ integers n_j, $j = 1, \ldots N$, such that:

$$z_i' = z_i - \sum_{j=1}^{N} \Delta_{ij} n_j \qquad \forall i \in V \tag{2.40}$$

This equivalence is said *equivalence under toppling*, and each equivalence class with respect to (2.40) contains one and only one recurrent configuration. One can associate to each configuration $z \in S^+$ a recurrent configuration $C[z]$ defining:

$$C[z] = \prod_i a_i^{z_i} z^* \tag{2.41}$$

where z^* is a given recurrent configuration. If z and z' are in the same equivalence class, then $C[z] = C[z']$, indeed we have that:

$$
\begin{aligned}
C[z'] &= \prod_i a_i^{z_i - \sum_j \Delta_{ij} n_j} z^* = \left(\prod_i a^{z_i}\right)\left(\prod_{ij} a^{\Delta_{ij} n_j}\right) z^* \\
&= \left(\prod_i a^{z_i}\right)\left(\prod_j \left(\prod_i a^{\Delta_{ij}}\right)^{n_j}\right) z^* = \prod_i a^{z_i} z^* = C[z_i].
\end{aligned}
\tag{2.42}
$$

Using the relation (2.40) two stable configurations can be equivalent under toppling. As a consequence of this equivalence relation and the existence of a unique recurrent representative for each equivalence class we will denote a class by $[z]$, being $[z] = [w]$ if z and w are equivalent under toppling. Furthermore the set of all configurations is a *superlattice* whose fundamental cell is the set $\mathbf{R}$, the rows of Δ are the principal vectors of the superlattice and $\det \Delta$ is the volume of the fundamental cell, that is the number of stable recurrent configurations $|\mathbf{R}|$.

We define *toppling invariants* as scalar functions over the space S_+ of all the configurations of the sandpile, such that their value is the same for configurations equivalent under toppling. Given the toppling matrix Δ for an N sites sandpile, we define N rational functions Q_i, $i \in \{1, \ldots, N\}$, as follows

$$Q_i(z) = \sum_j \Delta_{ij}^{-1} z_j \quad \mathrm{mod}\ 1 \tag{2.43}$$

It is straightforward to prove that the functions Q_i are toppling invariants, indeed a toppling at site k changes $z \equiv \{z_i\}$ into $z' \equiv \{z_i - \Delta_{ik}\}$, and the linearity of the Q_i's in the height variables allows to write:

$$Q_i(z') = Q_i(z) - \sum_j \Delta_{ij}^{-1} \Delta_{jk} = Q_i(z) \quad \mathrm{mod}\ 1 \tag{2.44}$$

These functions are rational-valued but they can be made integer-valued by multiplication upon an adequate integer. So these functions can be used to label the recurrent configurations. Thus the space of recurrent configurations $\mathbf{R}$ can be replaced by the set of N-uples $(Q_1, Q_2, \ldots, Q_N)$, but this labeling is generally overcomplete, they being not all independent.

It is desirable to isolate a minimal set of invariants, and this can be done for an arbitrary ASM using the classical theory of Smith normal form for integer matrices [12].

Any nonsingular $N \times N$ integer matrix Δ can be written in the form:

$$\Delta = ADB \tag{2.45}$$

where A and B are $N \times N$ integer matrices with determinant ± 1, and D is a diagonal matrix

$$D_{ij} = d_i \delta_{ij} \tag{2.46}$$

where the eigenvalues d_i are defined as follows:

1. d_i is a multiple of d_{i+1} for all $i = 1$ to $N - 1$
2. $d_i = e_{i-1}/e_i$ where e_i stands for the greatest common divisor of the determinants of all the $(N - i) \times (N - i)$ submatrices of Δ (note that $e_N = 1$)

The matrix D is uniquely determined by Δ but the matrices A and B are far from unique. The d_i are called the *elementary divisors* of Δ.

In terms of the decomposition (2.45), we define the set of scalar functions $I_i(z)$ by

$$I_i(z) = \sum_j (A^{-1})_{ij} z_j \quad \mathrm{mod}\, d_i \tag{2.47}$$

Due to the unimodularity of A (fact that guarantees the existence of an integer inverse matrix for A), these functions are integer-valued, and are toppling invariant, explicitly, given the equivalence under toppling relation (2.40), we have for $z \sim z'$:

$$I_i[z'] = \sum_j (A^{-1})_{ij} z_j - \sum_{jk} (A^{-1})_{ij} \Delta_{jk} n_k \tag{2.48}$$

$$= I_i[z] - \sum_{jk\ell m} (A^{-1})_{ij} A_{j\ell} D_{\ell m} B_{mk} n_k \tag{2.49}$$

$$= I_i[z] - \sum_{jk} D_{ij} B_{jk} n_k \tag{2.50}$$

$$= I_i[z] - d_i \sum_k B_{ik} n_k = I_i[z] \quad \mathrm{mod}\, d_i \tag{2.51}$$

Only the I_i for which $d_i \neq 1$ are nontrivial, and we note that this invariants are far from unique, because they are defined in the term of A which is not unique itself. The I_i's can also be written in term of the Q_i's as follows:

$$I_i[z] = \sum_j d_i B_{ij} Q_j[z] \tag{2.52}$$

We now show that the set of nontrivial invariants is always minimal and complete. Let g be the number of $d_i > 1$, we associate at each recurrent configuration a g-uple $(I_1, I_2, \ldots, I_g)$ where $0 \leq I_i < d_i$. The total number of distinct g-uple is $\prod_{i=1}^{g} d_i = |\mathcal{G}|$.

We first show that this mapping from the set of recurrent configurations to g-uples is one-to-one. Let us define operators e_i by the equation

$$e_i = \prod_{j=1}^{N} a_j^{A_{ji}} \qquad 1 \le i \le g \tag{2.53}$$

Acting on a fixed configuration $z^* = \{z_j\}$, e_i yields a new configuration, equivalent under toppling to the configuration $\{z_j + A_{ji}\}$. If the g-uple corresponding to z^* is $(I_1^*, I_2^*, \ldots, I_g^*)$, from (2.47) follows that $e_i z^*$ has toppling invariants $I_k = I_k^* + \delta_{ik}$. By operating with this operators $\{e_i\}$ sufficiently many times on z^*, all $|\mathcal{G}|$ values for the g-uple $(I_1, I_2, \ldots, I_g)$ are obtainable. Thus there is at least one recurrent configuration corresponding to any g-uple $(I_1, I_2, \ldots, I_g)$. As the total number of recurrent configurations equals the number of g-uples (2.22), we see that there is a one to one correspondence between the g-uples $(I_1, I_2, \ldots, I_g)$ and the recurrent configurations of the ASM.

To express the operators a_j in terms of e_i, we need to invert the transformation (2.53). This is easily seen to be:

$$a_j = \prod_{i=1}^{g} e_i^{(A^{-1})_{ij}} \qquad 1 \le j \le N \tag{2.54}$$

Thus the operators e_i generate the whole of $\mathcal{G}$. Since e_i acting on a configuration increases I_i by one, leaving the other invariants unchanged, and since I_i is only defined modulo d_i, we see that

$$e_i^{d_i} = I \quad \text{for } i \, 1 \text{ to } g \tag{2.55}$$

Note that the definition (2.53) makes sense for i between $g + 1$ and N, and implies relations among the a_j operators.

This shows that $\mathcal{G}$ has a canonical decomposition as a product of cyclic groups as in (2.39), with d_i's defined in (2.46). We thus have shown that the generators and the group structure of $\mathcal{G}$ can be entirely determined from its toppling matrix Δ, through its normal decomposition (2.45).

The invariants $\{I_i\}$ also provide a simple additive representation of the group $\mathcal{G}$. We define a binary operation of "addition" (denoted by $\oplus$) on the space of recurrent configurations by adding heights sitewise, and then allowing the resulting configuration to relax see (2.15). From the linearity of the I_i's in the height variables, and their invariance under toppling, it is clear that under this addition of configurations, the I_i's also simply add. Thus for any recurrent configurations z and w one has

$$I_i(z \oplus w) = I_i(z) + I_i(w) \mod d_i \tag{2.56}$$

The I_i's provide a complete labeling of **R**. There is a unique recurrent configuration, denoted by Id_r, for which all $I_i(Id_r)$ are zero. Also, each recurrent z has a unique inverse $-z$, also recurrent, and determined by $I_i(-z) = -I_i(z) \bmod d_i$. Therefore the addition $\oplus$ is a group law on **R**, with identity given by Id_r. M. Creutz first gave an algorithm to compute this configuration in [8, 9].

There is a one-to-one correspondence between recurrent configurations of ASM and elements of the group $\mathcal{G}$: we associate to the group element $g \in \mathcal{G}$, the recurrent configuration gId_r, and from (2.56) follows that for all $g, g' \in \mathcal{G}$

$$gId_r \oplus g'Id_r = (gg')Id_r \tag{2.57}$$

Thus the set of recurrent configurations with the operation $\oplus$ form a group which is isomorphic to the multiplicative group $\mathcal{G}$, result first proved in [8, 9]. The invariants $\{I_i\}$ provide a simple labeling of the recurrent configurations. Since a recurrent configuration can also be uniquely determined by its height variables $\{z_i\}$, the existence of forbidden configuration (2.30) in ASM's implies that this heights satisfy many inequality constraints.

2.5.2 Rank of $\mathcal{G}$ for a Rectangular Lattice

For a general toppling matrix Δ, it is difficult to say much more about the group structure of $\mathcal{G}$. To obtain some useful results we now consider the toppling matrix Δ of a finite $L_1 \times L_2$ bi-dimensional square lattice. In this framework is more convenient to label the sites not by a single index i running from 1 to $N = L_1 L_2$, but by two Cartesian coordinates (x, y), with $1 \leq x \leq L_1$ and $1 \leq y \leq L_2$. The toppling matrix is the discrete Laplacian as defined in (2.5), given by $\Delta(x, y; x, y) = 4$, $\Delta(x, y; x', y') = -1$ if the sites are nearest-neighbors (i.e. $|x - x'| + |y - y'| = 1$), and zero otherwise. We assume, without loss of generality, that $L_1 \geq L_2$. The relations (2.17) satisfied by operators $a(x, y)$, using the fact that the operators has an inverse on **R**, can be rewritten in the form

$$a(x + 1, y) = a^4(x, y)a^{-1}(x, y + 1)a^{-1}(x, y - 1)a^{-1}(x - 1, y) \tag{2.58}$$

where we adopt the convention that

$$a(x, 0) = a(x, L_2 + 1) = a(0, y) = a(L_1 + 1, y) = I \quad \forall x, y \tag{2.59}$$

The Eq. (2.58) can be recursively solved to express any operator $a(x, y)$ as a product of powers of $a(1, y)$. Therefore the group $\mathcal{G}$ can be generated by the L_2 operators $a(1, y)$. Denoting the rank of $\mathcal{G}$ (minimal number of generators) by g, this implies that

$$g \leq L_2 \tag{2.60}$$

In the special case of a linear chain, $L_2 = 1$, we see that $g = 1$, and thus $\mathcal{G}$ is cyclic. Equation (2.58) permits also to express $a(L_1 + 1, y)$ in term of powers of $a(1, y)$ say

$$a(L_1 + 1, y) = \prod_{y'} a(1, y')^{n_{yy'}} \tag{2.61}$$

where the $n_{yy'}$ are integers which depend on L_1 and L_2 and which can be eventually determined by solving the linear recurrence relation (2.58). The condition (2.59), $a(L_1 + 1, y) = I$ then leads to the closure relations

$$\prod_{y'=1}^{L_2} a(1, y')^{n_{yy'}} = I \quad \forall y = 1, \ldots, L_2 \tag{2.62}$$

The Eq. (2.62) give a presentation of $\mathcal{G}$, the structure of which can be determined from the normal form decomposition of the $L_2 \times L_2$ integer matrix $n_{yy'}$. This is considerably easier to handle that the normal form decomposition of the much larger matrix Δ needed for an arbitrary ASM. Even though this is a real computational improvement, the calculation for arbitrary L_1 is not trivial.

In the particular case of square-shaped lattice, where $L_1 = L_2 = L$, using the above algorithm is possible to find the structure of $\mathcal{G}$, and to prove that for an $L \times L$ square lattice we have

$$g = L \quad \text{for} \quad L_1 = L_2 = L \tag{2.63}$$

2.6 Generalized Toppling Rules

The two basic rules of the ASM are the *addition rule* and the *toppling rule*. The addition rule has a general formulation, and, in the identification of a Markov Chain, it is flexible because of the possibility to make different choices for the rates p_i, at which particles are added on each site. On the other hand the toppling rule is not in the most general formulation. Indeed the toppling rule has the form of a single-variable check, labeled by the site index i, $z_i < \bar{z}_i$. When the inequality fails causes an "instability" in the height profile. The heights are then relaxed with a constant shift $z \to z - \vec{\Delta}_i$ such that the total mass can only decrease. Some conditions on Δ ensure both the finiteness of the relaxation process, and the fact that there is no ambiguity in the "who topple first" in case of multiple violated disequalities at some intermediate steps.

Theorem 6.1 *Given an ASM on a graph $G = (E, V)$ and a toppling matrix Δ, if $\tilde{z}$ is unstable, consider the set S of sequences $(i_1, \ldots, i_{N(s)})$ such that $t_{i_{N(s)}} \ldots t_{i_2} t_{i_1} \tilde{z}$ is a valid sequence of topplings, and produces a stable configuration $z(s)$, some facts are true:*

(0) S is non-empty;

(1) $z(s) = z(s')$ for each $s, s' \in S$, i.e. the final stable configuration does not depend upon possible choices of who topples when;

(2) $N(s) = N(s') = N(\tilde{z})$ $\forall s, s' \in S$

(3) $\forall s, s' \in S$ $\exists \pi \in \mathcal{S}_{N(s)}$: $i_\alpha^{(s)} = i_{\pi(\alpha)}^{(s')}$ for $\alpha = 1, \ldots, N(\tilde{z})$, i.e. the toppling sequences differ only by a permutation.

Proof Here we prove *(3)*, given *(1)* and *(0)* As a restatement of *(3)*, we have that one can define a vector $\vec{n}(\tilde{z}) \in \mathbb{N}^{|V|}$ as the number of occurrence of each site in any of the sequence of S. Then, we have that the final configuration is

$$z = \tilde{z} + \Delta \cdot \vec{n} \tag{2.64}$$

The fact that $\vec{n}$ is unique is trivially proven. Indeed, as S is non-empty, we have a first candidate $\vec{n}_0$, and thus a solution of the non-homogeneous linear system (in $\vec{n}$)

$$\Delta \cdot \vec{n} = z - \tilde{z}.$$

If there was another solution $\vec{n}_1$, then we would have that $\vec{n}' = \vec{n}_1 - \vec{n}_0$ is a solution of the homogeneous system

$$\Delta \cdot \vec{n}' = 0.$$

But Δ is a square matrix of the form Laplacian+mass, such that the spectrum is all positive (we saw how it is a strictly-dissipative diffusion kernel), so it can not have non-zero vectors in its kernel. This proves the uniqueness of $\vec{n}$, i.e. *(3)*. But *(3)* is stronger than *(2)*, and the fact that $z = \tilde{z} + \Delta \cdot \vec{n}$ also implies *(1)*. So the theorem is proven.

The standard toppling rule can be shortly rewritten as:

$$\text{if } \exists i \in V \mid z_i \geq \bar{z}_i = \Delta_{ii} \implies z \to z + \vec{\Delta}_i \tag{2.65}$$

Pictorially, on a square lattice, we can draw the heights at a given site i and at its nearest-neighbors i_1, i_2, i_3, i_4 as

$$\begin{array}{c|c|c} & z_{i_1} & \\ \hline z_{i_4} & z_i & z_{i_2} \\ \hline & z_{i_3} & \end{array} \tag{2.66}$$

and an example of typical toppling is

$$\begin{array}{c|c|c} & 0 & \\ \hline 0 & 4 & 0 \\ \hline & 0 & \end{array} \longrightarrow \begin{array}{c|c|c} & 1 & \\ \hline 1 & 0 & 1 \\ \hline & 1 & \end{array} \tag{2.67}$$

where the initial value of z_i is equal to $\bar{z}_i = 4$, being the only site unstable, it requires the toppling shown in figure.

A straightforward generalization of this prescription is considering a site stable or unstable not just for "ultra-local" (i.e. single-site) properties -the overcoming of the site threshold- but also for local properties that depend on the heights at more than one sites. Similarly, we would have toppling rules $\vec{\Delta}_\alpha = \{\Delta_{\alpha j}\}_{j \in V}$ with more than a single positive entry. Still we want to preserve the *exchange* properties of toppling procedures which led to the abelianity of the *addition* operators a_i, and this could be in principle such a severe constraint that we could not find essentially any new possibility. As we show now, by direct construction, this is *not* the case. We can define some *cluster-toppling* rules, labeled by the subsets $I \subseteq V$ of the set of sites, instead that by a single site. For any subset I, we introduce the toppling rule

$$\text{if} \quad \forall i \in I \quad z_i \geq \sum_{\alpha \in I} \Delta_{i\alpha} \quad \Longrightarrow \quad z_k \to z_k - \sum_{\alpha \in I} \Delta_{\alpha k} \quad \forall k \in V \qquad (2.68)$$

These rules clearly define a new sandpile model, that, under some constraints on the choice of *toppling clusters set* $\mathcal{L} = \{I\}$, we will prove later to be abelian. Let us first address the simpler issue of checking for the finiteness of the space of configurations. It is trivial to see that if, for any site i, there is no single site set $\{i\} \in \mathcal{L}$, but only an arbitrary number of "large" clusters $I \in \mathcal{L}, |I| \geq 2, i \in I$, then all the configurations of height

$$z_i = n \in \mathbb{N}, \quad z_j = 0 \ j \neq i \qquad (2.69)$$

are allowed and stable, thus a necessary condition for having a finite set of stable configurations is

$$\{i\} \in \mathcal{L} \quad \forall i \in V \qquad (2.70)$$

this is also sufficient, as even in the standard ASM we have a number $\prod \bar{z}_i$ of stable configurations, and this number can only decrease when adding new toppling rules.

We define $\mathcal{L}_{std}$ the set of toppling cluster for the standard toppling rules, that is

$$\mathcal{L}_{std} = \{\{i\}_{i \in V}\} \qquad (2.71)$$

We ask now whether a given set $\mathcal{L}$ of cluster-toppling subsets give rise to an abelian sandpile. In the following we will denote with $I \setminus J$ the set of all the elements members of I but not members of J.

Theorem 6.2 *Given an ASM on a graph $G = (E, V)$, with a symmetric toppling matrix Δ, $\{\vec{\Delta}_I\}_{I \in \mathcal{L}}$ and $\mathcal{L} \supseteq \mathcal{L}_{std}$. A necessary and sufficient condition for the sandpile to be abelian is that*

$$\textit{each component of } I_{(1)} \setminus I_{(2)} \in \mathcal{L} \quad \forall I_{(1)}, I_{(2)} \in \mathcal{L} \qquad (2.72)$$

Proof Let us call $J = I_{(1)} \cap I_{(2)}$, then there are two cases:

(a) $J = \varnothing$
(b) $J \neq \varnothing$

In case (**a**) the compatibility is obvious, indeed if we make the toppling for $I_{(1)}$ then the heights in $I_{(2)}$ can only increase for the properties of the toppling matrix Δ. After the topplings also of the sites in $I_{(2)}$ have been done, the final height configuration will be

$$z'_k = z_k - \sum_{i \in I_{(1)}} \Delta_{ik} - \sum_{j \in I_{(2)}} \Delta_{jk} \tag{2.73}$$

this expression is clearly symmetric under the exchange of $I_{(1)}$ and $I_{(2)}$.

In case (**b**) we shortly recall the toppling rule for $I_{(1)}$ and $I_{(2)}$ (2.68)

$$\text{if } \forall i \in I_{(1)} \quad z_i \geq \sum_{\alpha \in I_{(1)}} \Delta_{i\alpha} \implies z_k \to z_k - \sum_{\alpha \in I_{(1)}} \Delta_{\alpha k} \quad \forall k \in V \tag{2.74a}$$

$$\text{if } \forall i \in I_{(2)} \quad z_i \geq \sum_{\alpha \in I_{(2)}} \Delta_{i\alpha} \implies z_k \to z_k - \sum_{\alpha \in I_{(2)}} \Delta_{\alpha k} \quad \forall k \in V \tag{2.74b}$$

now we note that we can split the sums in the contribution from J and the one from the remaining sites of each subset

$$\sum_{\alpha \in I_{(1)}} = \sum_{\alpha \in I_{(1)} \setminus J} + \sum_{\alpha \in J}$$

$$\sum_{\alpha \in I_{(2)}} = \sum_{\alpha \in I_{(2)} \setminus J} + \sum_{\alpha \in J}$$

Now we can topple first $I_{(1)}$ and so update the configuration $z \to z'$ as follows

$$z'_i = z_i + \sum_{\alpha \in I_{(1)} \setminus J} \Delta_{\alpha i} + \sum_{\alpha \in J} \Delta_{\alpha i} \quad \forall i \in V$$

At this point, for the updated configuration the following relations are valid

$$\forall i \in I_{(2)} \setminus J \quad z'_i = z_i - \sum_{\alpha \in I_{(1)} \setminus J} \Delta_{\alpha i} - \sum_{\alpha \in J} \Delta_{\alpha i} \tag{2.75a}$$

$$\geq \sum_{\alpha \in I_{(2)} \setminus J} \Delta_{i\alpha} + \sum_{\alpha \in J} \Delta_{i\alpha} - \sum_{\alpha \in I_{(1)} \setminus J} \Delta_{\alpha i} - \sum_{\alpha \in J} \Delta_{\alpha i} \geq \tag{2.75b}$$

$$\geq \sum_{\alpha \in I_{(2)} \setminus J} \Delta_{i\alpha} - \sum_{\alpha \in I_{(1)} \setminus J} \Delta_{\alpha i} \geq \tag{2.75c}$$

$$\geq \sum_{\alpha \in I_{(2)} \setminus J} \Delta_{i\alpha} \tag{2.75d}$$

in line (2.75b) we used the symmetry property of the toppling matrix to cancel out the second and the fourth terms, in line (2.75c) we used the property the off-diagonal elements Δ_{ij} to be negative or equal to zero to obtain the inequality in the last line, indeed if $A > B$ and $c_i \geq 0$, then $A + \sum c_i > B$ a *fortiori*.

In case it does not exist the toppling rule for $I_{(2)} \setminus J$, there exists a configuration of heights (the minimal heights such that both $I_{(1)}$ and $I_{(2)}$ are unstable) such that toppling first $I_{(1)}$ or $I_{(2)}$ leads immediately after a single toppling to two distinct stable configurations, so we see that necessary part of the theorem holds. As a consequence, as $\mathcal{L} \supseteq \mathcal{L}_{std}$, given $I \in \mathcal{L}$ we have that all the $I' \subseteq I$ are in $\mathcal{L}$, and thus all of its components. In particular disconnected I's are redundant, and we can restrict $\mathcal{L}$ to contain only connected clusters without loss of generality.

Conversely, if $I_{(2)} \setminus J \in \mathcal{L}$ (and $I_{(1)} \setminus J \in \mathcal{L}$ by symmetry), in the two "histories" in which we toppled $I_{(1)}$ or $I_{(2)}$, we can still topple $I_{(2)} \setminus J \in \mathcal{L}$ and $I_{(1)} \setminus J \in \mathcal{L}$ respectively and put them back on the same track, i.e.

$$t_{I_{(1)} \setminus J} t_{I_{(2)}} \equiv t_{I_{(2)} \setminus J} t_{I_{(1)}} \tag{2.76}$$

as operators when applied to configurations z such that both $I_{(1)}$ and $I_{(2)}$ are unstable.

We want also to produce a proof similar to that for standard toppling rule in theorem (6.1) for the cluster-toppling ASM.

Let suppose we have $G = (E, V)$, and the induced toppling matrix Δ, and a set $\mathcal{L}$ of connected subsets of V, with $\mathcal{L} \supseteq \mathcal{L}_{std} = \{\{i\}_{i \in V}\}$. Call $\vec{\Delta}_i = \{\Delta_{ij}\}_{j \in V}$, and $\vec{\Delta}_I = \sum_{i \in I} \vec{\Delta}_i$. A toppling t_I changes z into $z - \vec{\Delta}_I$.

Theorem 6.3 *If $\tilde{z}$ is unstable w.r.t. (2.68) given the framework above, consider the set S of sequences $s = (I_1, \ldots, I_{N(s)})$ such that $t_{I_{N(s)}} \ldots t_{I_2} t_{I_1} \tilde{z}$ is a valid sequence of topplings and produces a stable configuration $z(\tilde{z}; s)$. Some facts are true:*

(0) S is non-empty;
(1) $z(\tilde{z}; s) = z(\tilde{z}; s') \quad \forall s, s' \in S$;
(2) $\sum_{\alpha=1}^{N(s)} |I_\alpha^{(s)}| = \sum_{\alpha=1}^{N(s')} |I_\alpha^{(s')}| \quad \forall s, s' \in S$;
(3) defining $\vec{\chi}_I = \begin{cases} 1 & i \in I \\ 0 & i \notin I \end{cases}$, $\sum_{\alpha=1}^{N(s)} \vec{\chi}_{I_\alpha^{(s)}}$ is the same for all the sequences and is some vector $\vec{n}(\tilde{z})$

Proof (of *(3)* and *(2)* given *(1)* and *(0)*) Again, the final stable configuration is $z = \tilde{z} + \Delta \cdot \vec{n}$, and the uniqueness of $\vec{n}$ is proven along the same line as the proof for standard ASM. Then, as *(3)* is a strengthening of *(2)*, the theorem is proven. Remark however some qualitative difference with the simpler case of ordinary ASM: it can be that $N(s) \neq N(s')$, and s and s' do not differ simply by a permutation (e.g., in the relaxation of $\boxed{3}\boxed{4}\boxed{3}$ by $t_3 t_{12}$ or by $t_1 t_3 t_2$), and analogously the kernel

$$\sum_j n_I \vec{\Delta}_{Ij} = 0 \quad \forall I \in \mathcal{L}$$

is non-empty for $\mathcal{L} \supsetneq \mathcal{L}_{std}$, as Δ is rectangular (e.g. $n_{12} = a$, $n_1 = n_2 = -a$, $n_I = 0$ otherwise is a null vector of Δ). Only in the *basis* of $\mathcal{L}_{std}$ we have a unique solution, and of course the versor $\hat{e}_I$, in $\mathbb{Z}^{|\mathcal{L}|}$, reads $\vec{\chi}_I$ in this basis.

We present for clarity the example case in which the rule is defined for all the 2-clusters, *dimers*. In this case, given $G = (E, V)$ we have the set of toppling clusters

$$\mathcal{L} = \left\{ \{i, j\}_{ij \in E} \right\} \cup \left\{ \{i\}_{i \in V} \right\}, \tag{2.77}$$

and the general rule (2.68) becomes:

$$\text{if } \begin{cases} ij \in E \\ z_i \geq \Delta_{ii} + \Delta_{ij} \\ z_j \geq \Delta_{jj} + \Delta_{ij} \end{cases} \implies z_k \to z_k - \Delta_{ik} - \Delta_{jk} \quad \forall k \in V, \tag{2.78}$$

we can now pictorially draw on a square lattice the heights for a given cluster, formed by the sites i and j, and its nearest-neighbors $i_1, i_2, i_3, j_1, j_2, j_3$ as

$$\begin{array}{|c|c|c|c|} \hline & z_{i_1} & z_{j_1} & \\ \hline z_{i_3} & z_i & z_j & z_{j_2} \\ \hline & z_{i_2} & z_{j_3} & \\ \hline \end{array} \tag{2.79}$$

so a typical 2-cluster toppling is:

$$\begin{array}{|c|c|c|c|} \hline & 0 & 0 & \\ \hline 0 & 3 & 3 & 0 \\ \hline & 0 & 0 & \\ \hline \end{array} \longrightarrow \begin{array}{|c|c|c|c|} \hline & 1 & 1 & \\ \hline 1 & 0 & 0 & 1 \\ \hline & 1 & 1 & \\ \hline \end{array} \tag{2.80}$$

in which in the initial state both the sites i and j have height equal to $\bar{z}_i - 1 = 3$ and become unstable with respect to the (2.78), it is therefore necessary to topple the sites obtaining the final configuration. We note that doing a single cluster-toppling is the same as making two consecutive normal topplings, at condition that we permit negative height in the intermediate steps having forced the toppling in the case $z_i = 3$ when it is not necessary, in fact:

$$\begin{array}{|c|c|c|c|} \hline & 0 & 0 & \\ \hline 0 & 3 & 3 & 0 \\ \hline & 0 & 0 & \\ \hline \end{array} \longrightarrow \begin{array}{|c|c|c|c|} \hline & 1 & 0 & \\ \hline 1 & -1 & 4 & 0 \\ \hline & 1 & 0 & \\ \hline \end{array} \longrightarrow \begin{array}{|c|c|c|c|} \hline & 1 & 1 & \\ \hline 1 & 0 & 0 & 1 \\ \hline & 1 & 1 & \\ \hline \end{array} \tag{2.81}$$

and in the case $z_i \geq 4$ or $z_j \geq 4$ or both, the same result would have been obtained, any possible rule we choose to use, as proved in (6.2). This fact can be better understood recalling the relations (2.17) satisfied by the operators a_i and a_j, with i and j corresponding to the ones in (2.79):

$$a_i^4 = a_{i_1} a_{i_2} a_{i_3} a_j \tag{2.82a}$$

$$a_j^4 = a_{j_1} a_{j_2} a_{j_3} a_i \tag{2.82b}$$

If we restrict the attention on recurrent configurations where inverses of a_i exist it is possible to multiply side by side the two equalities obtaining:

$$a_i^4 a_j^4 = a_{i_1} a_{i_2} a_{i_3} a_j a_{j_1} a_{j_2} a_{j_3} a_i \tag{2.83}$$

and dividing (in group sense) each side by a_i and a_j we have the following equality:

$$a_i^3 a_j^3 = a_{i_1} a_{i_2} a_{i_3} a_{j_1} a_{j_2} a_{j_3} \tag{2.84}$$

that is the same of (2.82) for the cluster toppling rule. This rule easily generalizes for arbitrary subsets of V. This permit us to state that the different toppling rules we have introduced bring to the same group presentation (and then to the same group structure) for the abelian group associated to the recurrent configurations of the ASM.

We recall now that for the model with the standard toppling rule we have an easy characterization for the subsets F of the graph that are forbidden subconfiguration (FSC), that is:

$$\text{if } \forall i \in F \quad z_i < \sum_{\substack{j \neq i \\ j \in F}} \left(-\Delta_{ij}\right) \quad \Longrightarrow \quad F \text{ is a FSC} \tag{2.85}$$

As obvious with the new rules just introduced, some forbidden subconfigurations of the standard ASM can become reachable by adding sand and toppling, e.g. the simplest forbidden configuration in the case of standard toppling rules, $\boxed{0\,|\,0}$, is easily reachable if we allow the dimer-cluster toppling rule, in fact it can simply turn out as result of the basic dimer-cluster toppling $\boxed{3\,|\,3} \rightarrow \boxed{0\,|\,0}$. It is also possible to characterize the forbidden subconfigurations with respect to a given I-toppling rule, we have:

$$\text{if } \forall I \subseteq F \quad \sum_{j=1}^{k} z_{i_j} < \sum_{\substack{j \in F \setminus I \\ \ell \in I}} \left(-\Delta_{\ell j}\right) \quad \Longrightarrow \quad F \text{ is a FSC} \tag{2.86}$$

this yielding to the possibility that a transient configuration with respect to a $\mathcal{L}'$ toppling rule becomes recurrent for a $\mathcal{L}''$ toppling rule, with $\mathcal{L}'' \supset \mathcal{L}'$.

On the other hand some configurations stable with respect to a $\mathcal{L}'$ toppling rule become unstable if we allow $\mathcal{L}'$ to increase up to $\mathcal{L}''$, e.g. the basic unstable configuration $\boxed{3\,|\,3}$ for the dimer-toppling rule is perfectly stable in the framework of toppling only for $z_i \geq \bar{z}_i$. Moreover the fact that the group structure of the associated group remains unchanged under the addition of the new toppling rules, yields the number of group elements $g \in \mathcal{G}$ to be the same in the two cases, this forces the number of recurrent configurations to be the same, as the order of group associated, i.e. as many stable recurrent becomes unstable, as transient become recurrent, for each enlargement of $\mathcal{L}$.

In this framework, we see how, remaining unchanged the number of recurrent configurations and growing up the number of unstable configurations, since the set of $\mathcal{L}$-stable configurations becomes a subset of the original set of stable configurations $S = \{0, 1, 2, 3\}^{|V|}$, in some sense the transient configurations that become allowed must correspond to some newly unstable configurations. This kind of symmetry between the two sets, yields to suppose the existence of a bijection between unstable configurations for $\mathcal{L}''$ toppling rules and transient configurations for $\mathcal{L}'$ toppling rules, with $\mathcal{L}'' \supset \mathcal{L}'$. This procedure of enlarging the set of unstable configuration, and at the same time to shrink the set of transient configuration yields to the possibility to completely suppress the set of the transient configurations and to have that the recurrent configurations become all the stable configurations. This situation is reached by letting

$$\mathcal{L} = \{\text{all the subsets of } V\} \tag{2.87}$$

An interesting example is the lattice 3×1 in which the number of configurations is not huge and we can directly check this statement.

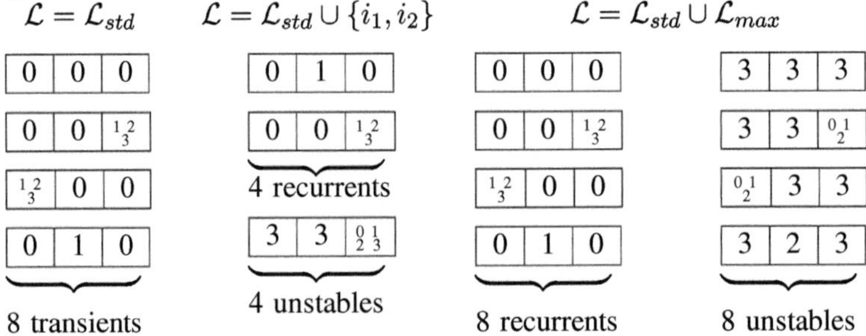

References

1. P. Bak, C. Tang, K. Wiesenfeld, Self-organized criticality: an explanation of the 1/f noise. Phys. Rev. Lett. **59**, 381–384 (1987)
2. D. Dhar, Theoretical studies of self-organized criticality. Physica A **369**, 29–70 (2006)
3. E. Ivashkevich, V. Priezzhev, Introduction to the sandpile model. Physica A **254**, 97–116 (1998)
4. R. Meester, F. Redig, D. Znamenski, The abelian sandpile: a mathematical introduction. Markov process Relat. Fields **7**, 509 (2001)
5. F. Redig, Mathematical aspects of the abelian sandpile model, in Mathematical statistical physics École d'été de physique des houches session LXXXIII, vol. 83 of Les Houches, (Elsevier, 2006), pp. 657–729
6. D. Dhar, Self-organized critical state of sandpile automaton models. Phys. Rev. Lett. **64**, 1613–1616 (1990)
7. S.S. Manna, Large-scale simulation of avalanche cluster distribution in sand pile model. J. Stat. Phys. **59**, 509–521 (1990). doi:10.1007/BF01015580
8. M. Creutz, Abelian sandpiles. Comp. Phys. **5**, 198–203 (1991)

9. M. Creutz, Abelian sandpiles, Nucl. Phys. B (Proc. Suppl.), **20**, 758–761 (1991)
10. S.N. Majumdar, D. Dhar, Equivalence between the abelian sandpile model and the $q \to 0$ limit of the potts model. Physica A **185**, 129–145 (1992)
11. E.R. Speer, Asymmetric abeiian sandpile models. J. Stat. Phys. **71**, 61–74 (1993). doi:10.1007/BF01048088
12. N. Jacobson, *Basic Algebra*, 2nd edn. (W.H. Freeman and Company, New York, 1985)

Chapter 3
Algebraic Structure

ASM as a Monoid

In this chapter we give a further insight in the algebraic structure of the ASM. We start in the first section recalling some notions on a number of known facts but using the formalism we will use further on. We will introduce from the beginning new operators, defined by means of *antitoppling rules*, that we will call $a_i^\dagger$'s and are the *symmetric* counterpart of the operators a_i. Such operators were already introduced by Manna et al. in [1] as *hole addition* then followed by *hole avalanches* in order to reach a stable configuration. Also inverse avalanches has been introduced before, in [2], to get back to the recurrent configuration after the deletion of a particle in a configuration of the ASM.

The transition monoid corresponding to the dynamics ruled by the *toppling rules* alone, in the set of stable configurations, is *abelian*; this is a property which seems at the basis of our understanding of the model. By including also *antitoppling rules*, and the operators $a_i^\dagger$'s, we introduce and investigate a larger monoid, which is *not abelian* anymore. We prove a number of algebraic properties of this monoid, and describe their practical implications on the emerging structures of the model.

We shows in the last section the structure of Markov chains dynamics involving both a_i and $a_i^\dagger$, elucidating the role played by the relations among the operators in the evolution of the dynamics.

Part of the results here presented have been submitted to be published in [3].

3.1 The Extended Configuration Space

3.1.1 Algebraic Formalism

Let the integer n be the *size* of the system. It is often useful to think at the system as a graph with n sites, and the set of toppling rules in terms of the adjacency structure of the graph, thus we will call *sites* the indices $i \in V \equiv \{1, \ldots, n\}$. Consider vectors $z \in \mathbb{Z}^n$, where we will use the partial ordering $\preceq$ such that $u \preceq v$ if $u_i \leq v_i$ for each

G. Paoletti, *Deterministic Abelian Sandpile Models and Patterns*,
Springer Theses, DOI: 10.1007/978-3-319-01204-9_3,
© Springer International Publishing Switzerland 2014

$i \in V$. We will define the *positive cone* Ω as the subset of $w \in \mathbb{Z}^n$ such that $w \succeq 0$, where 0 has vanishing entries for all i.

An abelian sandpile $\mathcal{A} = \mathcal{A}(\Delta, \overline{z}, \underline{z})$ is identified by a triple $(\Delta, \overline{z}, \underline{z})$, that we now describe. The vectors $\overline{z}$ and $\underline{z}$ are the collection of *upper-* and *lower-thresholds*, $\{\overline{z}_i\}$ and $\{\underline{z}_i\}$ respectively, and are constrained to the condition $\overline{z}_i - \underline{z}_i > 0$ for all i. Define the spaces

$$S_+ = \{z \in \mathbb{Z}^n \mid z \succeq \underline{z}\} = \bigotimes_{i=1}^{n} \{\underline{z}_i, \underline{z}_i + 1, \ldots\}; \tag{3.1}$$

$$S_- = \{z \in \mathbb{Z}^n \mid z \preceq \overline{z}\} = \bigotimes_{i=1}^{n} \{\ldots, \overline{z}_i - 1, \overline{z}_i\}; \tag{3.2}$$

$$S = \{z \in \mathbb{Z}^n \mid \underline{z} \preceq z \preceq \overline{z}\} = \bigotimes_{i=1}^{n} \{\underline{z}_i, \ldots, \overline{z}_i\} = S_+ \cap S_-. \tag{3.3}$$

Thus S_+ and $-S_-$ are translations of Ω, while S is a multidimensional interval, and has finite cardinality.

We also have a $n \times n$ *toppling matrix* Δ, with integer entries, that should satisfy $\overline{z}_i - \underline{z}_i + 1 \geq \Delta_{ii} > 0$, and $\Delta_{ij} \leq 0$ for $i \neq j$. We say that the sandpile is *tight* if $\overline{z}_i - \underline{z}_i + 1 = \Delta_{ii}$ for all i. Here we restrict our study at tight sandpiles.[1]

We further require *dissipativity*, that is $b_i^- = \sum_j \Delta_{ij} \geq 0$. As seen below, b_i^- is the amount of mass that *leaves* the system after a toppling at i. The requirement that the toppling matrix is *irreducible*[2] ensures that the avalanches are finite (and that $\det \Delta > 0$). For future utility, we also define $b_j^+ = \sum_i \Delta_{ij}$, which is the difference between the amount of mass that leave the site j in a toppling and the mass that is there added if all other sites would make a toppling. A site j where $b_i^+ < 0$ is said to be *greedy* or *selfish*. Clearly $\sum_i b_i^- = \sum_i b_i^+$, so also the b_i^+'s are, on average, higher than zero, however positivity on the b_i^+'s is not implied directly by the positivity of the b_i^-'s. We require $b_i^+ \geq 0$ as an extra condition, whose utility will be clear only in the following. A sandpile is said to be *unoriented* if $\Delta = \Delta^{\mathrm{T}}$ (and thus $b^- = b^+$).

The above conditions complete the list of constraints characterizing valid triples $(\Delta, \overline{z}, \underline{z})$. The special case of the BTW sandpile corresponds to Δ being the discretized Laplacian on the square lattice, $\Delta_{ii} = 4$, $\Delta_{ij} = -1$ if $d(i, j) = 1$ and $\Delta_{ij} = 0$ if $d(i, j) > 1$, where $d(i, j)$ is the Euclidean distance between the sites i and j, and $\overline{z}_i = 3$, $\underline{z}_i = 0$ for all i.

The matrix Δ is the collection of the toppling rules, and is conveniently seen as a set of vectors $\vec{\Delta}_i = \{\Delta_{ij}\}_{1 \leq j \leq n}$. Denote by t_i the action of a toppling at i. If such a toppling occurs, the configuration z is transformed according to

$$t_i z = z - \vec{\Delta}_i. \tag{3.4}$$

---

[1] For *tight* sandpiles S_+ and S_- are equal to $S_+ = \bigotimes_{i=1}^{n} \{\overline{z}_i - \Delta_{ii} + 1, \overline{z}_i - \Delta_{ii} + 2, \ldots\}$ and $S_- = \bigotimes_{i=1}^{n} \{\ldots, \underline{z}_i + \Delta_{ii} - 2, \underline{z}_i + \Delta_{ii} - 1\}$.

[2] This means that for every j_0 there exists a sequence $(j_0, j_1, \ldots, j_\ell)$ such that $\Delta_{j_a\, j_{a+1}} < 0$ for all $0 \leq a < \ell$, and $b_{j_\ell}^- > 0$.

A site i is *positively-unstable* (or just *unstable*) if $z_i > \bar{z}_i$. In this case, and only in this case, a toppling can be performed at i. Note that, after the toppling, it is still $z_i \geq \underline{z}_i$ (more precisely, $z_i > \bar{z}_i - \Delta_{ii}$), while z_j's for $j \neq i$ have not decreased, thus the topplings leave stable the space S_+. The *relaxation operator* $\mathcal{R}$ is the map from S_+ to S, coinciding with the identity on S, that associates to a configuration $z \in S_+$ the unique configuration $\mathcal{R}(z) \in S$ resulting from the application of topplings at unstable sites. Unicity relies on the abelianity of the toppling rules.

A site i is *negatively-unstable* if $z_i < \underline{z}_i$. In this case, an *antitoppling* can be performed at i, with the rule

$$t_i^\dagger z = z + \vec{\Delta}_i \,. \tag{3.5}$$

We use deliberately the symbol $t_i^\dagger$ instead of t_i^{-1} because, although the effect of the linear transformation (3.5) is just the inverse of the effect of (3.4), the *if* conditions for applicability of the two operators are different.

Now antitopplings leave stable the spaces S_-. The *antirelaxation operator* $\mathcal{R}^\dagger$ is the map from S_- to S, coinciding with the identity on S, that associates to a configuration $z \in S_-$ the unique configuration $\mathcal{R}^\dagger(z) \in S$ resulting from the application of antitopplings at negatively-unstable sites. As a matter of fact, the involution

$$\iota : \quad z \to \bar{z} + \underline{z} - z \tag{3.6}$$

exchanges the role of operators with and without the $\dagger$ suffix, i.e., for the operators above and all the others introduced later on, we have $A^\dagger(z) \equiv \iota A(\iota z)$.

Note that, for a configuration $z \in \mathbb{Z}^n$, we cannot exchange in general the ordering of topplings and antitopplings (i.e., if i and j are respectively positively- and negatively-unstable for z, j might be stable for $t_i z$, and i might be stable for $t_j^\dagger z$). Consistently, we do *not* define any relaxation-like operator from $\mathbb{Z}^n$ to S.

Remark, however, that the definition of $\mathcal{R}$ can be trivially extended in order to map unambiguously $\mathbb{Z}^n$ to S_-, by letting it produce a toppling only on unstable sites, and, analogously, $\mathcal{R}^\dagger$ to $\mathbb{Z}^n$ to S_+ (and both $\mathcal{R}^\dagger \mathcal{R}$ and $\mathcal{R} \mathcal{R}^\dagger$ map $\mathbb{Z}^n$ to S, but $\mathcal{R}^\dagger(\mathcal{R}(z)) \neq \mathcal{R}(\mathcal{R}^\dagger(z))$ in general).

On the space $\mathbb{Z}^n$ we can of course take linear combinations, and define the *sum* of two configurations, $z + w$, and the *multiplication by scalars* $k \in \mathbb{Z}$, kz. For generic $\bar{z}$ and $\underline{z}$ these operations do not leave stable any of the subsets $S_\pm$ and S. But sum and difference, $z + w$ and $z - w$, are binary operations in the following spaces:

$$z + w : \qquad\qquad S_+ \times \Omega \to S_+ \,; \tag{3.7}$$

$$z - w : \qquad\qquad S_- \times \Omega \to S_- \,. \tag{3.8}$$

Thus, in particular, the two maps $\mathcal{R}(z + w)$ and $\mathcal{R}^\dagger(z - w)$ can act from $S \times \Omega$ into S. This suggests to give a special name and symbol to the simplest family of these binary operations, seen as operators on S. Call e_i the canonical basis of $\mathbb{Z}^n$, i.e., $(e_i)_j = \delta_{i,j}$. Define the operators $\hat{a}_i$ such that $\hat{a}_i z = z + e_i$, and introduce the operators of *sand addition* and *removal*

$$a_i = \mathcal{R}\hat{a}_i \; ; \qquad\qquad\qquad a_i^\dagger = \mathcal{R}^\dagger \hat{a}_i^{-1} \; . \qquad\qquad (3.9)$$

The a_i's commute among themselves, i.e., for every z, $a_i a_j z = a_j a_i z = \mathcal{R}(z + e_i + e_j)$. Similarly, the $a_i^\dagger$'s commute among themselves. More generally, for any $z \in \mathbb{Z}^n$ and $w \in \Omega$, $\mathcal{R}(\mathcal{R}(z) + w) = \mathcal{R}(z + w)$, and $\mathcal{R}^\dagger(\mathcal{R}^\dagger(z) - w) = \mathcal{R}^\dagger(z - w)$, this implying

$$\mathcal{R}(z + w) = \left(\prod_{i=1}^n (a_i)^{w_i}\right) z \; ; \qquad \mathcal{R}^\dagger(z - w) = \left(\prod_{i=1}^n (a_i^\dagger)^{w_i}\right) z \; . \qquad (3.10)$$

This can be seen by induction in i, as $\mathcal{R}(z + (w' + e_i)) = \mathcal{R}(\mathcal{R}(z + e_i) + w') = \mathcal{R}((a_i z) + w'))$. For later convenience, for $w \in \Omega$, we introduce the shortcuts

$$a_w = \prod_{i=1}^n (a_i)^{w_i} = \mathcal{R}(\cdot + w) \; ; \qquad a_w^\dagger = \prod_{i=1}^n (a_i^\dagger)^{w_i} = \mathcal{R}^\dagger(\cdot - w) \; ; \qquad (3.11)$$

(note that the order in the products does not matter).

These properties have an important consequence on the structure of the Markov Chain dynamics under which ASM has been studied when introduced by Bak et al. At each integer time t a site $i(t)$ is chosen at random, and $z(t + 1) = a_{i(t)} z(t)$. As long as we are interested in the configuration $z(t_\text{fin})$ for a unique final time t_fin, for a given initial state $z(0)$, it is not necessary to follow the entire evolution $z(t)$, for $0 \le t \le t_\text{fin}$, but it is enough to take the vector $w = \sum_t e_{i(t)}$, and evaluate $\mathcal{R}(z(0) + w)$. Thus, the final result of the time evolution depends on the set $\{i(t)\}_{0 \le t < t_\text{max}}$ of moves at all times, in a way which is invariant under permutations.

Note however that, similarly to topplings with antitopplings, also the operators of sand addition and removal do not commute among themselves, i.e., $a_i\, a_j^\dagger z \ne a_j^\dagger a_i\, z$ in general, even for $z \in S$. Therefore, in a Markov process involving both a_i's and $a_i^\dagger$'s, in order to know the final configuration, it is necessary to follow the full trace of the time evolution. We will briefly describe and investigate a dynamics of this kind, for the sandpile on a square lattice, in Sect. 3.4. Furthermore, see [1] for a first extensive investigation of a dynamics in this family.

Beside the commutativity relations $a_i a_j = a_j a_i$ (and conjugated ones), there is a collection of n relations, encoded by the toppling matrix: for any i, when acting on configurations such that $z_i > \bar{z}_i - \Delta_{ii}$, (respectively, such that $z_i < \underline{z}_i + \Delta_{ii}$), we have

$$a_i^{\Delta_{ii}} = \prod_{j \ne i} a_j^{-\Delta_{ij}} \; ; \qquad\qquad (a_i^\dagger)^{\Delta_{ii}} = \prod_{j \ne i} (a_j^\dagger)^{-\Delta_{ij}} \; . \qquad (3.12)$$

This because $t_i \hat{a}_i^{\Delta_{ii}} = \prod_{j \ne i} \hat{a}_j^{-\Delta_{ij}}$ on such configurations, and the site i is certainly positively-unstable after the application of $\hat{a}_i^{\Delta_{ii}}$.

In the previous paragraphs we only considered sums $z + w$, in which one of the two configurations is taken from a space among S, or $S_\pm$, and the other one from Ω, or other spaces with no dependence from $\overline{z}$ and $\underline{z}$.

In such a situation, we have a clear covariance of the notations under an overall translation of the coordinates. Indeed, for any $r \in \mathbb{Z}^n$, under the map $z \to z + r$, we have an isomorphism between the sandpile $\mathcal{A}(\Delta, \overline{z}, \underline{z})$ and $\mathcal{A}(\Delta, \overline{z} + r, \underline{z} + r)$. We call a *gauge invariance* of the model the covariance explicitated above, and a *gauge fixing* any special choice of offset vector r.

As always, a gauge fixing reduces the apparent number of parameters in the model. Some special choices simplify the notations in certain contexts. For example, we can set $\underline{z} = 0$, so that $S_+ \equiv \Omega$, that we call the $\underline{z} = 0$ *gauge*, or $\overline{z}_i = \Delta_{ii} - 1$, that we call the *positive-cone gauge*.

In particular, the positive-cone gauge is the most natural one in the Abelian Sandpile in which antitoppling are not considered, as in this case the parameters $\underline{z}$ do not play any role. Conversely, the covariant formalism is the only formulation that does not break explicitly the involution symmetry implied by (3.6).

3.1.2 Further Aspects of the Theory

There exists a natural equivalence relation on vectors $z \in \mathbb{Z}^n$, that partitions this set into $\det \Delta$ classes, which are affine subspaces of $\mathbb{Z}^n$, all isomorphic under translation. This notion was first introduced and studied in [4]. We recall here only briefly the easiest facts.

We say that $z \sim w$ if there exists $T \in \mathbb{Z}^n$ such that $z - w = T\Delta$. In particular, as $z - t_i z = -(z - t_i^\dagger z) = \vec{\Delta}_i = e_i \Delta$, we have $z \sim \mathcal{R}(z) \sim \mathcal{R}^\dagger(z)$. Analogously, as $\hat{a}_i^{\Delta_{ii}} z - \left(\prod_{j \neq i} \hat{a}_j^{-\Delta_{ij}}\right) z = -e_i \Delta$, we get $a_i^{\Delta_{ii}} z \sim \left(\prod_{j \neq i} a_j^{-\Delta_{ij}}\right) z$ for all i, as it should at the light of (3.12). The dissipativity condition on the toppling matrix ensure that $\det \Delta \neq 0$. Thus Δ^{-1} exists, and it is evident that $z \sim w$ if and only if $z\Delta^{-1} - w\Delta^{-1} \in \mathbb{Z}^n$. So, the fractional parts $Q_i^{(\mathrm{frac})}(z) = (z\Delta^{-1})_i - \lfloor (z\Delta^{-1})_i \rfloor$, called the *charges* of the configuration z, completely identify the equivalence class of z. As a corollary, $Q_i^{(\mathrm{frac})}(z) = Q_i^{(\mathrm{frac})}(t_i z) = Q_i^{(\mathrm{frac})}(t_i^\dagger z)$ (when t_i or $t_i^\dagger$ are applicable to z). For future convenience, we also define $Q(z) = (z\Delta^{-1})$, so that $Q_i^{(\mathrm{frac})} = Q_i - \lfloor Q_i \rfloor$.

The set S of stable configurations is divided into the two subsets of *stable transient* and *stable recurrent* configurations, $S = T \cup \mathbf{R}$. Several equivalent characterizations of recurrency exist, some of which extend naturally to S_+, and even to the full space $\mathbb{Z}^n$ (this is also our choice).

In particular, we give three definitions, all valid in $\mathbb{Z}^n$. Under all definitions, a configuration is *transient* if it is not recurrent.

Definition 3.1 *A configuration z is* recurrent by identity test *if there exists a permutation $\sigma \in \mathfrak{S}_n$ such that $t_{\sigma(n)} \cdots t_{\sigma(2)} t_{\sigma(1)}(z + b^+) = z$ is a valid toppling sequence.*

Definition 3.2 *A configuration z is* recurrent by toppling covering *if there exists a configuration u, such that $z = t_{i_k} \cdots t_{i_1} u$ is a valid toppling sequence, and at least one toppling is performed at each site.*

Definition 3.3 *A configuration z is* recurrent by absence of FSC's *if, for every set I of sites, there exists $i \in I$ such that $z_i > \overline{z}_i - \sum_{j \in I} \Delta_{ji}$.*

The reasonings of the following paragraphs will prove, among other things, that these three definitions are equivalent.

First we note that a configuration recurrent by identity test is also recurrent by toppling covering (one can take $u = z + b^+$).

Note that, if $u = \mathcal{R}(v)$ and v is recurrent by toppling covering, also u is recurrent by toppling covering. Furthermore, as after a toppling t_i one has $z_i > \overline{z}_i - \Delta_{ii}$, all the recurrent configurations, either by toppling covering or by identity test, are in fact contained in S_+. This last reasoning is a first example of *forbidden sub-configuration* (FSC), whose generalization involves more than one site at a time. For $z \in \mathbb{Z}^n$, and $I \subseteq [n]$, define $z|_I$ as the restriction to the components z_i with index $i \in I$.

For a set I, define the vector $f_{\max}(I) \in \mathbb{Z}^I$ as

$$\left(f_{\max}(I) \right)_i = \overline{z}_i - \sum_{j \in I} \Delta_{ji} \, . \tag{3.13}$$

We say that the pair $(I, z|_I)$ is a *forbidden sub-configuration* for z if $z \preceq f_{\max}(I)$. It is straightforward to recognize that z is recurrent for absence of FSC's if and only if, for all I, $z|_I \not\preceq f_{\max}(I)$, thus legitimating the terminology.

A collection of the pairs $(I, f_{\max}(I))$ with smallest $|I|$ in the BTW sandpile is as follows:

$$\boxed{-1} \qquad \boxed{0\,|\,0} \qquad \genfrac{}{}{0pt}{}{\boxed{0}}{\boxed{0}} \qquad \boxed{0\,|\,1\,|\,0} \qquad \genfrac{}{}{0pt}{}{\boxed{0}}{\boxed{1\,|\,0}} \;\; \cdots \;\; \boxed{0\,|\,1\,|\,1\,|\,0} \;\; \cdots \;\; \genfrac{}{}{0pt}{}{\boxed{1\,|\,1}}{\boxed{1\,|\,1}} \;\; \cdots$$

The connection between the definitions of recurrent by toppling covering and by absence of FSC's is given by the following statement, slightly more general that what would suffice at this purpose

Proposition 3.1 *Let $u \in S'_+$ and $v = t_{i_k} t_{i_{k-1}} \cdots t_{i_1} u$. Define $A = \bigcup_{1 \leq a \leq k} \{i_a\}$. Then, for all B such that $|B \smallsetminus A| \leq 1$, $v|_B \not\preceq f_{\max}(B)$.*

This proposition implies as a corollary (for $|B \smallsetminus A| = 0$) that configurations which are recurrent by toppling covering are also recurrent by absence of FSC's. The case $|B \smallsetminus A| = 1$ also emerges naturally from the proof.

Proof The claim $v|_B \not\preceq f_{\max}(B)$ can be restated as the existence of $s \in B$ such that $v_s > f_{\max}(B)_s$. We will produce a valid choice for s.

If $|B \smallsetminus A| = 1$, choose as s the only site in B and not in A. Let $u' = u$ and $\tau(s) = 0$ in this case. Otherwise, for all $i \in B$, call $\tau(i)$ the maximum $1 \leq a \leq k$ such that $i_a = i$, then choose as s the index realising the minimum of $\tau(i)$, i.e., the site of B that has performed its last toppling more far in the past. Call $u' = t_{i_{\tau(s)}} \cdots t_{i_1} u$, the configuration obtained after the last toppling in s.

Note that, as $u \in S'_+$ and this space is stable under topplings, $u'_s \geq \overline{z}_s - \Delta_{ss} + 1$. In the remaining part of the avalanche, no more topplings occur at s. Furthermore, all the other sites $j \in B$ do topple at least once, at $a = \tau(j)$. Thus we have

$$v_s = u'_s - \sum_{a=\tau(s)+1}^{k} \Delta_{i_a s} \geq \overline{z}_s - \Delta_{ss} + 1 - \sum_{a=\tau(s)+1}^{k} \Delta_{i_a s}$$

$$= \overline{z}_s - \sum_{i \in B} \Delta_{is} + 1 - \sum_{\substack{\tau(s)<a\leq k \\ a \notin \{\tau(j)\}_{j \in B}}} \Delta_{i_a s} \geq \overline{z}_s - \sum_{i \in B} \Delta_{is} + 1 . \tag{3.14}$$

The comparison with the definition (3.13) of $f_{\max}(B)$ allows to conclude. $\square$

For a sandpile $\mathcal{A}(\Delta, \overline{z}, \underline{z})$, call $\mathcal{A}|_I$ the sandpile described by the toppling matrix $\Delta_{I,I}$ the principal minor of Δ corresponding to the rows and columns in I, and by the threshold vectors $\overline{z}|_I$ and $\underline{z}|_I$. Also call $b^+(I)$ the vector b^+ associated to $\Delta_{I,I}$. The observation that ultimately allows to relate the definitions of recurrency by absence of FSC's and by identity test is the fact that

$$f_{\max}(I) + b^+(I) = \overline{z}|_I . \tag{3.15}$$

A first remark in this direction is that, if $I = [n]$ is not a FSC for z, we can at least start the avalanche in the definition of recurrent by identity test, i.e., we have at least one site $i = \sigma(1)$ which is unstable. Indeed, we have at least one i such that $z_i > f_{\max}([n])_i$. Then, by (3.15), $(z + b^+)_i > f_{\max}([n])_i + b_i^+ = \overline{z}_i$.

As (3.15) holds for any set I, the reasoning above works for any restricted sandpile $\mathcal{A}|_I$. Suppose to have a configuration $z^{(I)}$ which is recurrent in $\mathcal{A}|_I$ by absence of FSC's. Then we have at least one site $i \in I$ which is unstable, because we have at least one i such that $z_i^{(I)} > f_{\max}(I)_i$, and $(z^{(I)} + b^+(I))_i > f_{\max}(I)_i + b^+(I)_i = \overline{z}_i$.

This allows to construct an induction. Let $I' = I \smallsetminus i$, $v^{(I')} = (t_i z^{(I)})|_{I'}$ and $z^{(I')} = \mathcal{R}(v^{(I')})$. For $j \in I'$, $v_j^{(I')} = z_j^{(I)} - \Delta_{ij} \geq z_j^{(I)}$. Remark that the definition of $f_{\max}(J)$ is the same on any restricted sandpile $\mathcal{A}_I$ with $I \supseteq J$. Thus, as $z^{(I)}$ is recurrent for absence of FSC's, and this property is preserved under relaxation (by Proposition 3.1), also $z^{(I')}$ is recurrent by absence of FSC's, on $\mathcal{A}|_{I'}$. This, together with equation (3.15), gives the induction step.

In summary, we can perform the complete avalanche $t_{\sigma(n)} \cdots t_{\sigma(1)}$, algorithmically, by initialising $z^{(0)} = z + b^+$ and $I_0 = [n]$, and, for $a = 0, \ldots, n-1$, i_a is any site i at which $z^{(a)}$ is unstable (which is proven to exist by the reasoning above), $I_{a+1} = I_a \smallsetminus i_a$, and $z^{(a+1)} = \mathcal{R}(t_{i_a} z^{(a)})$.

This completes the equivalence of our three definitions of recurrent configurations, thus from now on we will omit to specify the defining property.

A directed graph can be associated to a sandpile, such that $-\Delta_{ij}$ directed edges connect i to j. This completely encodes the off-diagonal part of Δ. The remaining parameters, in particular b^+ and b^-, can be encoded through directed edges incoming from, our outgoing to, a special 'sink' vertex. This can be done unambiguously if $b_i^+ \geq 0$ for all i. In the undirected case, $\Delta = \Delta^{\mathrm{T}}$, a bijection due to Majumdar and Dhar [5], called *Burning Test*, relates stable recurrent configurations to spanning trees. A generalization, called *Script Algorithm* and due to Speer [6] relates stable recurrent configurations to directed spanning trees rooted at the sink vertex. Through Kirchhoff Matrix-Tree Theorem, one thus gets in this case that the number of stable recurrent configurations is $\det \Delta$.

The configuration $p = \bar{z} + 1 - \mathcal{R}(\bar{z} + 1)$ has two interesting properties: $p_i \geq 1$ for all i, and $p \sim 0$. This implies that, for every $z \in \mathbb{Z}^n$, the iteration of the map $z \to \mathcal{R}(z + p)$ must reach a fixed point, in $\mathbf{R}$, the subset of S containing recurrent configurations (this would be true also with b^+ instead of p, but slightly harder to prove). Indeed, calling $c = \max_i (\bar{z}_i - z_i + 1)$, as $\mathcal{R}(\mathcal{R}(\cdots \mathcal{R}(z + p) \cdots + p) + p) = \mathcal{R}(z + c\,p)$, and $z + c\,p$ is unstable at all sites, $\mathcal{R}(z + c\,p)$ is both stable and recurrent (by toppling covering), so it must be in $\mathbf{R}$.

This reasoning proves that each equivalence class has at least one representative in $\mathbf{R}$. In the case $b^+ \in \Omega$, as both the cardinality of $\mathbf{R}$ and the number of classes are $\det \Delta$, each equivalence class must have a unique representative in $\mathbf{R}$. In particular, the representative in $\mathbf{R}$ of 0 is called the *recurrent identity*, and we denote it with the symbol Id_{r}. This configuration can be found as the fixed point of the map $z \to \mathcal{R}(z + b^+)$, started from 0 (see [7, 8]), or of the map $z \to \mathcal{R}(z + p)$, or, under the mild assumption that $\bar{z} \in \Omega$, more directly, with no need of iterations, by the relation

$$Id_{\mathrm{r}} = \mathcal{R}(\bar{z} + (\bar{z} - \mathcal{R}(\bar{z} + \bar{z}))) . \tag{3.16}$$

(introduced in [9]). In fact, as $\bar{z} - \mathcal{R}(\bar{z} + \bar{z}) \in \Omega$, $\bar{z} + (\bar{z} - \mathcal{R}(\bar{z} + \bar{z}))$ is recurrent, thus its relaxation is in $\mathbf{R}$.

In the $\underline{z} = 0$ gauge, the operation $u \oplus v := \mathcal{R}(u + v)$ sends $S \times S$ into S, and thus defines a semigroup on this space. Furthermore, this operation also sends $S \times \mathbf{R} \to \mathbf{R}$, $\mathbf{R} \times S \to \mathbf{R}$, and, as a corollary, $\mathbf{R} \times \mathbf{R} \to \mathbf{R}$.

The charges behave linearly under this operation: $Q(u \oplus v) - Q(u) - Q(v) \in \mathbb{Z}^n$. The unicity of representatives in $\mathbf{R}$ of the equivalence classes allows then to construct inverses, for the action on this space, and thus to promote $\oplus$ to a group action on $\mathbf{R}$, of which Id_{r} is the group identity. This structure was first introduced by Creutz [7, 8], and then investigated in several papers [4, 9, 10].

The covariance of the notations allows to define the operation $\oplus$ in general. We have

$$u \oplus v := \mathcal{R}(u + v - \underline{z}) ; \qquad u \oplus^{\dagger} v := \mathcal{R}^{\dagger}(u + v - \bar{z}) . \tag{3.17}$$

Note however that now the charges behave in an affine way, and only the translated versions, $Q'(u) = Q(u) - Q(\underline{z})$ and $Q''(u) = Q(u) - Q(\bar{z})$ respectively for the two operations, have no offset. In particular, the two groups induced by $\oplus$ and $\oplus^{\dagger}$, as well as the two sets $\mathbf{R}$ and $\mathbf{R}^{\dagger}$, are isomorphic but not element-wise coincident, as the only natural bijection among the two makes use of the involution ι.

Call $\mathcal{M} = \mathcal{M}[a_i, a_i^\dagger]$ the transition monoid generated by the a_i's and $a_i^\dagger$'s acting on our set of configurations (see [11] for an introduction to the theory). A generic element in $\mathcal{M}$ has the form

$$A = a_{i_1^1}^\dagger \cdots a_{i_{\ell(1)}^1}^\dagger \, a_{i_1^2} \cdots a_{i_{\ell(2)}^2} \, a_{i_1^3}^\dagger \cdots a_{i_{\ell(3)}^3}^\dagger \quad \cdots \quad a_{i_1^{2k}} \cdots a_{i_{\ell(2k)}^{2k}} \ . \tag{3.18}$$

These monomials, for different sets of indices, are not all different among each others (we recall that $\mathcal{M}$, seen as a transition monoid acting on $\mathbb{Z}^n$, has all the relations $a_i a_j = a_j a_i$, and $a_i^\dagger a_j^\dagger = a_j^\dagger a_i^\dagger$, while acting on S has also all the relations (3.12). Define

$$w(A) = -e_{i_1^1} \cdots - e_{i_{\ell(1)}^1} + e_{i_1^2} \cdots + e_{i_{\ell(2)}^2} - e_{i_1^3} \cdots - e_{i_{\ell(3)}^3} \quad \cdots \quad + e_{i_1^{2k}} \cdots + e_{i_{\ell(2k)}^{2k}} \ ,$$
$$\tag{3.19}$$

which is thus independent on the ordering of the operators within A (that, recall, is composed of *non-commutative* operators). For A and A' two operators of the form above then we have $\vec{w}(AA') = \vec{w}(A) + \vec{w}(A')$, while, analyzing the charges defined for the configurations, we have for any $z \in \mathbb{Z}^n$, $Q^{(\mathrm{frac})}(Az) - Q^{(\mathrm{frac})}(w(A)) - Q^{(\mathrm{frac})}(z) \in \mathbb{Z}^n$. Thus, we have a collection of sub-monoids $\mathcal{M}_H$ of $\mathcal{M}$ associated to a given sandpile $\mathcal{A}$, which are in correspondence with the subgroups $H \leq G$ of the group of translations on the torus, $G \cong \mathbb{Z}_{d_1} \times \mathbb{Z}_{d_2} \times \cdots \times \mathbb{Z}_{d_g}$ describing the structure of the set **R** of stable recurrent configurations.

In particular, the smallest of these sub-monoids, that we call $\mathcal{M}_0$, is in correspondence with the trivial subgroup on the torus composed of the identity element alone, and contains all and only the operators A with $w(A) \sim 0$. This is the sub-monoid of operators that leave stable the equivalence classes, i.e., $Az \sim z$.

The simplest operators in $\mathcal{M}_0$ are the set of $a_i^\dagger a_i$, and $a_i a_i^\dagger$, for any i. Slightly less simple combinations are $a_w^\dagger a_w$ and $a_w a_w^\dagger$, for generic $w \in \Omega$. Indeed, the fact that these operators are elements in $\mathcal{M}_0$ has been one of the motivations behind the study of relations as in Theorems 3.4 and 3.5. The study of a dynamics involving the operators $a_i^\dagger a_i$ is suggested in Sect. 3.4.

We have also other sub-monoids, that do not follow under the characterization above. In particular, of course, we have the two commutative sub-monoids $\mathcal{M}^+$ and $\mathcal{M}^-$, generated by the sand addition operators alone, $\{a_i\}$, or by the sand removal operators alone, $\{a_i^\dagger\}$, respectively, together with their sub-monoids $\mathcal{M}_H^\pm$, again in correspondence with subgroups of the torus. The monoids $\mathcal{M}_0^+$ and $\mathcal{M}_0^-$ contain the identities of the sandpile (e.g., if $b_i^+ \geq 0$ for all i, $a_{b^+} \in \mathcal{M}_0^+$). Remark however how in general all monomials in $\mathcal{M}_0^\pm$ different from the identity have 'large degree' (e.g., of order L for the BTW sandpile on a $L \times L$ domain), while in $\mathcal{M}_0$ the simplest non-trivial elements, $a_i^\dagger a_i$, have degree 2.

3.2 Statement of Results

Two new theoretical results are extensively required for the analysis of interesting dynamics. The first one is the following

Theorem 3.4 *For every* $i \in V$, *acting on* S_+,

$$a_i \, a_i^\dagger a_i = a_i \; ; \tag{3.20}$$

and, acting on S_-,

$$a_i^\dagger a_i \, a_i^\dagger = a_i^\dagger \; . \tag{3.21}$$

Proof The two equations are related by the involution, so we only prove (3.20). I.e., for all $z \in S_+, a_i \, a_i^\dagger a_i \, z = a_i \, z$. First of all, as, for $z \in S_+$ and $w \in \Omega, a_w z = a_w \mathcal{R}(z)$, we can restrict our attention to $z \in S$. If $z_i < \bar{z}_i$ we have $a_i^\dagger a_i \, z = z$, and our relation follows. If $z_i = \bar{z}_i$, the avalanche due to the action of a_i performs at least one toppling at i. By Proposition 3.1, $y = a_i z$ has no FSC's with $I = \{i\}$ or $I = \{i, j\}$. That is, either $y_i > \bar{z}_i - \Delta_{ii}$, or $y_i = \bar{z}_i - \Delta_{ii}$ and, for all $j \neq i$, $y_j \geq \bar{z}_j - \Delta_{jj} - \Delta_{ij}$. By direct inspection, in the first case $a_i \, a_i^\dagger y = \hat{a}_i \hat{a}_i^{-1} y = y$ and in the second case $a_i \, a_i^\dagger y = t_i \, \hat{a}_i t_i^\dagger \hat{a}_i^{-1} y = y$. In both cases, our relation follows. $\qquad \square$

As a matter of fact, Theorem 3.4 is in fact the tip of an iceberg of a much more general family of identities, with operators a_i and $a_i^\dagger$ replaced by a_w and $a_w^\dagger$, for $w \in \Omega$, that will be discussed in the next Section. Let us note here how, acting on S_+, the identity $a_w a_w^\dagger a_w = a_w$ holds for simple reasons if w is *recurrent*, due to the fact that $a_w a_w^\dagger a_w z \sim a_w z$, both sides of the equation are stable recurrent, and there exists a unique stable recurrent representative in each equivalence class.

An immediate corollary of Theorem 3.4 is the following

Corollary 3.1 *For every* $i \in V$, *acting on* S, $\Pi_i \equiv a_i^\dagger a_i$ *and* $\Pi_i^\dagger \equiv a_i \, a_i^\dagger$ *are idempotents, i.e.,* $\Pi_i^2 = \Pi_i$, *and* $\Pi_i^{\dagger 2} = \Pi_i^\dagger$.

Indeed, the set $S = S_+ \cap S_-$ is left stable by the action of the monoid. It is enough to multiply equation (3.20) by $a_i^\dagger$, from the left or from the right respectively for the two claims, or, alternatively, multiply (3.21) by a_i , from the right or from the left respectively.

The simplicity of Corollary 3.1 may suggest that abelianity is restored at the level of these idempotent combinations, Π_i. This is not the case. No pairs of distinct operators in the set $\{a_1^\dagger a_1, \ldots, a_n^\dagger a_n, a_1 a_1^\dagger, \ldots, a_n a_n^\dagger\}$ commute with each other, in general. Nonetheless, a few interesting facts are found.

For a finite set $I \subseteq V$, call $\mathcal{N}_I = \{a_i^\dagger a_i\}_{i \in I}$. For $X \subseteq \mathbb{Z}^n$, call $\mathcal{N}_I[X]$ the set of $y \in \mathbb{Z}^n$ such there exists a configuration $x \in X$ and a finite sequence $(i_1, \ldots, i_k)$ of elements in I such that $a_{i_k}^\dagger a_{i_k} a_{i_{k-1}}^\dagger a_{i_{k-1}} \cdots a_{i_1}^\dagger a_{i_1} x = y$, that is $\mathcal{N}_I[X]$ is the set of possible images of X under the action of products of operators in $\mathcal{N}_I$.

The second new theoretical result of this paper is

Theorem 3.5 *Consider a sandpile $\mathcal{A}(\Delta, \bar{z}, \underline{z})$ such that $\Delta = \Delta^{\mathrm{T}}$ and $\Delta_{ij} \in \{0, -1\}$ for $i \neq j$. For any $I \subseteq V$, and any $z \in S_+$, there exists a unique state $y(z, I)$ in $\mathcal{N}_I[\{z\}]$ such that $a_i^\dagger a_i\, y = y$ for all $i \in I$. For any state $x \in \mathcal{N}_I[\{z\}]$, we also have $y \in \mathcal{N}_I[\{x\}]$.*

Thus, this theorem shows that certain collections of idempotents have a well-characterised set of common fixed points. The portion of this set accessible from any configuration z has cardinality exactly 1.

Furthermore, the statement of this theorem can be translated in terms of Markov Chains. Consider a Markov Chain in which the initial state is $z(0) = z$, and at each time t an element $a_{i_t}^\dagger a_{i_t} \in \mathcal{N}_I$ is chosen (with non-zero probabilities for all elements), and $z(t+1) = a_{i_t}^\dagger a_{i_t} z(t)$. Then the theorem states that this Markov Chain is absorbent, on an unique state y, and in particular, no matter the evolution up to some time t, y is still accessible from $z(t)$ (and in fact it *will* be reached at some time). A dynamics of this kind will be described in Sect. 3.4.

The theorem above will be proven in Sect. 3.3. Furthermore, the state $y(z, I)$ will be shown to have a further characterization, in terms of a multitoppling Abelian Sandpile associated to the original system.

3.2.1 Generalization of Theorem 3.4

As anticipated in the previous Section, Theorem 3.4 is only a particular case of a wider class of identities. Our aim is to prove the Theorem 3.7 which states

$$a_w a_w^\dagger a_w z = a_w z . \qquad (3.22)$$

In order to achieve this result, we start with a lemma.

Let $z \in S_+$, and define $\mathcal{T}(z) = \{T \mid T \in \Omega, z - T\Delta \in S\}$. This set is non-empty, as it contains the vector $T^*(z)$ defined as $\mathcal{R}(z) = z - T^*\Delta$.

Lemma 3.2 *For all $T \in \mathcal{T}(z)$, $T \succeq T^*$.*

Proof This is a consequence of abelianity. Call $X^{(1)}$ the set of unstable sites of $z \equiv z^{(1)}$, and $T^{(1)}$ as $T_i^{(1)} = 1$ if $i \in X^{(1)}$ and 0 otherwise. For $z - T\Delta \in S$, one needs $T \succeq T^{(1)}$. Call $z^{(2)} = z^{(1)} - T^{(1)}\Delta$. Note that $z^{(2)} \in S_+$. Clearly, $T \in \mathcal{T}(z^{(1)})$ if and only if $T - T^{(1)} \in \mathcal{T}(z^{(2)})$. Repeating the resoning above for $z^{(2)}$ leads that, for $z - T\Delta \in S$, one needs $T \succeq T^{(1)} + T^{(2)}$. This analysis corresponds to the implementation of the relaxation operator "in parallel". Iterating, up to when the avalanche stops, leads to $T \succeq T^{(1)} + T^{(2)} + \cdots = T^*$, as was to be proven. $\square$

Lemma 3.3 *If $z \in \mathbf{R}$, $T \in \Omega$, and $y = z + T\Delta \in S$, then $T = 0$ and $y = z$.*

Proof Assume by absurd that $T \succ 0$. Call $m = \max_i(T_i) \geq 1$ and X_m the set of sites realising the maximum. We prove that X_m is a FSC for z. Indeed, for $i \in X_m$,

$$f_{\max}(X_m)_i = \bar{z}_i - \sum_{j \in X_m} \Delta_{ji} , \qquad (3.23)$$

while

$$z_i = y_i - \sum_j T_j \Delta_{ji} = y_i - m \sum_{j \in X_m} \Delta_{ji} - \sum_{j \notin X_m} T_j \Delta_{ji}$$

$$\leq \bar{z}_i - m \sum_{j \in X_m} \Delta_{ji} - \sum_{j \notin X_m} (m-1)\Delta_{ji} = \bar{z}_i - \sum_{j \in X_m} \Delta_{ji} - (m-1)b_i^+ .$$

$$(3.24)$$

Where the last equality comes by adding and subtracting $\sum_{j \in X_m} \Delta_{ji}$ and substituting b_i^+ where necessary. Then we know that $m \geq 1$ and $b_i^+ \geq 0$ and comparing with (3.23), we conclude. $\qquad\square$

Theorem 3.6 *For $w \in \Omega$ and $z \in \mathbf{R}$,*

$$a_w a_w^\dagger z = z . \qquad (3.25)$$

Proof Let $T^{(1)}$ and $T^{(2)}$ the vectors such that $y = a_w^\dagger z = z - w + T^{(1)}\Delta$ and $x = a_w y = y - T^{(2)}\Delta = z + (T^{(1)} - T^{(2)})\Delta$. As $T^{(1)} \in \mathcal{T}(y + w)$ (because $(y + w) - T^{(1)}\Delta = z \in S$), and by definition $T^{(2)} = T^*(y + w)$, by Lemma 3.2 we have that $T^{(2)} \preceq T^{(1)}$, i.e., $T^{(1)} - T^{(2)} \in \Omega$. Finally, as $z \in \mathbf{R}$, by Lemma 3.3 we have that $x = z$, as was to be proven. $\qquad\square$

Now we give three main theorems, the first of them is the natural generalization of Theorem 3.4 whose proof is the aim of this section. We first prove their mutual equivalence. Then, we will deduce one of them from the Theorem 3.6 above.

Theorem 3.7 *For $w \in \Omega$ and $z \in S_+$,*

$$a_w a_w^\dagger a_w z = a_w z . \qquad (3.26)$$

Theorem 3.8 *For $u, v \in \Omega$ and $z \in S_+$,*

$$a_u \, a_v a_v^\dagger a_v \, z = a_v a_v^\dagger a_v \, a_u \, z . \qquad (3.27)$$

Theorem 3.9 *For $u, v \in \Omega$ and $z \in S_+$, calling $w = u + v$,*

$$a_u a_u^\dagger a_u \, a_v a_v^\dagger a_v \, z = a_w a_w^\dagger a_w \, z . \qquad (3.28)$$

Proof Theorem 3.7 implies Theorem 3.8, by abelianity of the a_i's.

Consider Theorem 3.8, specialized to $z = 0$,

$$a_u\, a_v a_v^\dagger a_v\, 0 = a_v a_v^\dagger a_v\, a_u\, 0\,, \tag{3.29}$$

which is trivially rewritten as

$$a_u\, v = a_v a_v^\dagger a_v\, u\,. \tag{3.30}$$

and recalling that $a_v\, u = a_u\, v$

$$a_v\, u = a_v a_v^\dagger a_v\, u\,. \tag{3.31}$$

which is Theorem 3.7

Theorem 3.7 implies Theorem 3.9, again by abelianity of the a_i's.

Theorem 3.6 implies Theorem 3.7, again specializing to the case of $z = 0$. Indeed

$$a_u a_u^\dagger a_u\, a_v a_v^\dagger a_v\, 0 = a_u a_u^\dagger a_u\, v\,. \tag{3.32}$$

while

$$a_w a_w^\dagger a_w\, 0 = w = u + v = a_u\, v\,. \tag{3.33}$$

And this concludes the equivalences. $\square$

We observe that Theorem (3.7) has already been proved in the previous Section in the special case $w = e_i$, and there is also pointed out how it is trivially valid in the case of $w \in \mathbf{R}$ and $s \in S^+$.

Proof of the Theorem 3.7 Let us call $z^{(1)} = a_w z$, and $T^{(1)}$ the corresponding toppling vector, $z^{(1)} = z + w - T^{(1)}\Delta$.

Let us call X_+ and X_0 the sets of sites for which $T^{(1)} > 0$ and $T^{(1)} = 0$, respectively, and call w^+ and w^0 the vectors

$$w_i^+ = \begin{cases} w_i & i \in X_+ \\ 0 & i \in X_0 \end{cases} \qquad\qquad w_i^0 = \begin{cases} 0 & i \in X_+ \\ w_i & i \in X_0 \end{cases} \tag{3.34}$$

We observe that $z^{(1)}$ has no FSC's contained in X_+.

The theorem is equivalent to

$$a_w a_w^\dagger z^{(1)} = z^{(1)}\,. \tag{3.35}$$

For all i,

$$z_i^{(1)} = z_i + w_i - \sum_j T_j^{(1)}\Delta_{ji}\,, \tag{3.36}$$

and in particular, if $i \in X_0$, all summands $-T_j^{(1)}\Delta_{ji}$ are non-negative. We write (3.35) as

$$a_w a^\dagger_{w+} a^\dagger_{w0} z^{(1)} = z^{(1)} .$$ (3.37)

The action of $a^\dagger_{w0}$ on $z^{(1)}$ is trivial, thanks to the observation above. Calling $z^{(2)} = z^{(1)} - w^0$, the action of a_{w0} on $z^{(2)}$ is trivial for the same reason, and we have

$$a_{w0} a_{w+} a^\dagger_{w+} z^{(2)} = a_{w0} z^{(2)} .$$ (3.38)

We will prove the stronger

$$a_{w+} a^\dagger_{w+} z^{(2)} = z^{(2)} .$$ (3.39)

Define $z^{(3)} = z^{(2)} - w^+ + T^{(2)} \Delta$, and $z^{(4)} = z^{(3)} + w^+ - T^{(3)} \Delta$. Using Lemma 3.2, through a reasoning analogous to the one done in the proof of Theorem 3.6, we get $T^{(3)} \preceq T^{(2)} \preceq T^{(1)}$, and the theorem is proven if we find $T^{(3)} = T^{(2)}$. The inequalities above imply in particular that $T^{(2)}_j = T^{(3)}_j = 0$ if $j \in X_0$. Thus, the identity (3.39) is equivalent to its restriction to the induced sandpile on the set X_+. But, as $z^{(2)}|_{X_+} = z^{(1)}|_{X_+}$ is recurrent, the restriction of Eq. (3.39) is in fact the statement of Theorem 3.6. This concludes the proof.   □

Corollary 3.4 *The following relations hold and are equivalent for $u, v, w \in \Omega$ and $z \in S_-$; here $w = u + v$*
For $w \in \Omega$ and $z \in S_-$,

$$a^\dagger_u a_u a^\dagger_u z = a^\dagger_u z ;$$ (3.40)

$$a^\dagger_u a^\dagger_v a_v a^\dagger_v z = a^\dagger_v a_v a^\dagger_v a^\dagger_u z ;$$ (3.41)

$$a^\dagger_u a_u a^\dagger_u a^\dagger_v a_v a^\dagger_v z = a^\dagger_w a_w a^\dagger_w z .$$ (3.42)

Proof The relations are related by the involution, 3.6, the the statement of Theorems 3.7, 3.8 and 3.9, so they are trivially proved.   □

3.3 Multitopplings in Abelian Sandpiles

We recall in this Section the theory of *multitoppling* already introduced in Sect. 2.6 with a notation consistent to the rest of this chapter in order to obtain a proof for Theorem 3.5. Consider an ordinary ASM $\mathcal{A}(\Delta, \bar{z}, z)$. For simplicity of notations, we set in the positive-cone gauge, so that site i is unstable if $z_i \geq \Delta_{ii}$.

For a non-empty set $I \subseteq [n]$, call $\vec{\Delta}_I = \sum_{i \in I} \vec{\Delta}_i$. A multitoppling operator t_I can be associated to a set I. First of all, z is unstable for toppling I if $z_j \geq (\Delta_I)_j$ for all $j \in I$ (note how, in the positive-cone gauge, this coincides with the ordinary definition when $|I| = 1$). Then, if the configuration is unstable, it is legitimate to perform the toppling $t_I z = z - \vec{\Delta}_I$. Also note that the dissipativity condition on Δ implies that, for all $j \in I$, $(\Delta_I)_j \geq 0$, while for all $j \notin I$, $(\Delta_I)_j \leq 0$.

Consider a collection $\mathcal{L}$ of non-empty subsets of $[n]$. The interest in multitoppling rules for the Abelian Sandpile Model is in the following fact

Proposition 3.2 *Suppose that, for every I, $J \in \mathcal{L}$, the sets $I' = I \setminus J$ and $J' = J \setminus I$ are either empty or in $\mathcal{L}$. Then, if z is unstable for both I and J, $t_I z$ is unstable for J' and $t_J z$ is unstable for I'.*

As clearly $t_{J'} t_I z = t_{I'} t_J z$, it easily follows

Corollary 3.5 *In the conditions of Proposition 3.2, the operator $\mathcal{R}$ is unambiguous.*

Antitopplings $t_I^{\dagger}$ and antirelaxation $\mathcal{R}^{\dagger}$ are defined just as in the ordinary case, e.g., through the involution ι, which is still defined as in (3.6). A configuration is stable if no toppling or antitoppling can occur (this coincides with the definition of S in the ordinary case). Note that, if $\mathcal{L}$ does not contain the atomic set $\{i\}$, S is either empty or of infinite cardinality (because, if $z \in S$, also $z + c\,\vec{e}_i \in S$ for any $c \in \mathbb{Z}$). In order to exclude this pathological case, we will assume in the following that $\mathcal{L}$ includes the set $\mathcal{L}_0 = \left\{\{i\}\right\}_{i \in [n]}$ of all atomic subsets, i.e., single-site topplings. In this case, as stability w.r.t. $\mathcal{L}$ is a more severe requirement than stability w.r.t. $\mathcal{L}' \subset \mathcal{L}$, we have that S is a subset of the set S_0 of stable configurations in the associated sandpile with only single-site topplings, and thus of finite cardinality. As we will see later on, for any set $\mathcal{L}$ as in Proposition 3.2 the set S is non-empty, and actually contains a set isomorphic to $\mathbf{R}$, thus it has cardinality bounded below by $|R_0| = \det \Delta$, an above by $|S_0| = \prod_i (\bar{z}_i - \underline{z}_i + 1)$.

If we require both that $\mathcal{L} \supseteq \mathcal{L}_0$, and satisfies the hypotheses of Proposition 3.3, we get that $\mathcal{L}$ is a down set in the lattice of subsets, that is, for all $I \in \mathcal{L}$ and $H \subseteq \mathcal{L}$ non-empty, also $H \in \mathcal{L}$ (this is trivially seen: with notations as in Proposition 3.2, take I and $J = I \setminus H$).

For a multitoppling sandpile $\mathcal{A} = \mathcal{A}(\mathcal{L})$, we will call $\mathcal{A}_0 = \mathcal{A}(\mathcal{L}_0)$ the associated single-site toppling sandpile.

The concept of recurrent configuration is rewritten in the context of multitoppling rules. The various alternate definitions are modified.

Definition 3.4 *A configuration z is* recurrent by identity test *if there exists an ordered sequence $(I_1, \ldots, I_k)$ of subsets of $[n]$, constituting a partition of $[n]$ (i.e., for all $i \subset [n]$ there exists a unique a such that $i \in I_a$), such that $t_{I_k} \cdots t_{I_2} t_{I_1} (z + \vec{b}^+) = z$ is a valid toppling sequence.*

Definition 3.5 *A configuration z is* recurrent by toppling covering *if there exists a configuration u, such that $z = t_{I_k} \cdots t_{I_1} u$ is a valid toppling sequence, and each site i is contained in at least one of the I_a's.*

Definition 3.6 *A configuration z is* recurrent by absence of FSC's *if, for every set I of sites, there exists $J \in \mathcal{L}$ with $L = I \cap J \neq \emptyset$, and $z_j \geq \sum_{i \in I \setminus L} \Delta_{ij}$ for all $j \in L$.*

All the reasonings are the direct generalization of the ones already given in Sect. 3.1.2. We just report here the appropriate modifications in Proposition 3.1 (recall that in our gauge $\Omega \equiv S'_{+}$, and is left stable by the topplings).

Proposition 3.3 *Let $u \in \Omega$ and $v = t_{I_k} t_{I_{k-1}} \cdots t_{I_1} u$. Define $A = \bigcup_{1 \leq a \leq k} I_a$. For any set B, there exists a non-empty set $H \subseteq B$, such that $B \subseteq H \cup A$, and, for all $s \in H$, $v_s > -\sum_{i \in (A \cap B) \setminus H} \Delta_{is}$.*

Proof We will produce explicitly a valid choice of H. If $B \supseteq A$, let $H = B \setminus A$ and $u' = u$. Note that $(B \cap A) \setminus H = B \cap A$ in this case. If $B \not\supseteq A$, for all $i \in B$, call $\tau(i)$ the maximum $1 \leq a \leq k$ such that $i \in I_a$, then call $\tau = \max_i \tau(i)$, and $J = I_{\tau(i)}$, i.e., the multitoppling that covered any portion of B more far in the past. Call $u' = t_{I_\tau} \cdots t_{I_1} u$, the configuration obtained after this last multitoppling. Note that all the entries of u' are non-negative. In the remaining part of the avalanche, for some non-empty set $H \subseteq J$, no more topplings occur. Conversely, all the sites $j \in (A \cap B) \setminus H$ do topple at least once (possibly in a multitoppling event). Thus we have, for each $s \in H$,

$$v_s \geq - \sum_{i \in (B \cap A) \setminus H} \Delta_{is} \,, \tag{3.43}$$

as was to be proven. □

We should modify the concept of forbidden sub-configuration along the same lines. We want to produce pairs (I, f) such that, if z is recurrent by toppling covering, $z|_I \neq f$. For a given I, a vector f has the property above if, for all $J \in \mathcal{L}$, $J \subseteq I$, $f|_J \preceq -\sum_{k \in I \setminus J} \vec{\Delta}_k|_J$. Note that the fact that $\mathcal{L}$ is a down set has been used to restrict the set of J's to analyse.

A collection of the forbidden pairs (I, f) with smallest $|I|$, such that no other f' exists with $f' \succ f$ and (I, f') a forbidden pair, in the BTW sandpile with multitoppling rules on all pairs of adjacent sites, is as follows:

$$\begin{array}{|c|}\hline -1 \\ \hline \end{array} \qquad \begin{array}{|c|c|c|}\hline 0 & 0 & 0 \\ \hline \end{array} \qquad \begin{array}{|c|c|}\hline 0 & \multicolumn{1}{|c}{} \\ \hline 0 & 0 \\ \hline \end{array} \quad \ldots \quad \begin{array}{|c|c|}\hline 0 & 1 \\ \hline 1 & 0 \\ \hline \end{array} \qquad \begin{array}{|c|c|}\hline 1 & 0 \\ \hline 0 & 1 \\ \hline \end{array} \quad \ldots$$

Note that, at difference with the single-site sandpile, in general there is not a unique $f_{\max}(I)$ such that (I, f) is a forbidden pair if and only if $f \preceq f_{\max}(I)$.

At the level of the monoid $\mathcal{M}[a_i, a_i^\dagger]$, we have some extra relations associated to multitoppling rules. For example, consider the action on S_+. While in the single-site sandpile we have relations (3.12), in the multitoppling case we also have relations of the form, for each $I \in \mathcal{L}$,

$$\prod_{j \in I} a_j^{(\Delta_I)_j} = \prod_{j \notin I} a_j^{-(\Delta_I)_j} \,. \tag{3.44}$$

Note that multiplying left-hand and right-hand sides of relations (3.12) for all $j \in I$ we would have got a weaker relation, in which the two sides of (3.44) are multiplied by

$$\prod_{\substack{i \in I \\ j \in I \setminus i}} a_j^{-\Delta_{ij}} , \tag{3.45}$$

(recall that at the level of the monoid it is not legitimate to take inverses of a_i's, and that, in the equations above, all the exponents are indeed non-negative).

At the level of the equivalence relation $\sim$, and thus of charges $\vec{Q}(z)$, nothing changes. In particular, it is easy to see that $u \sim v$ in $\mathcal{A}$ if and only if $u \sim v$ in $\mathcal{A}_0$. This also leads to the fact that there is exactly one stable recurrent configuration per equivalence class, and that $\oplus$ defines a group structure over $\mathbf{R}$, just as in the single-site toppling sandpile $\mathcal{A}_0 = \mathcal{A}(\mathcal{L}_0)$ associated to the multitoppling sandpile $\mathcal{A} = \mathcal{A}(\mathcal{L})$.

Note however that the set $\mathbf{R}$ is *different* from the set $\mathbf{R}_0$ of stable recurrent configurations for $\mathcal{A}_0$, and the fact that they have the same cardinality results from subtle compensations between stable/unstable and recurrent/transient configurations.

A natural bijection between $\mathbf{R}$ and $\mathbf{R}_0$, preserving the group structure, is obtained by associating to $z \in \mathbf{R}_0$ the configuration $\mathcal{R}(z) \in \mathbf{R}$, where $\mathcal{R}$ is the complete (multitoppling) relaxation.

In the case of a tight sandpile, to which we restricted in this chapter, the set $\mathbf{R}^\dagger$, containing the conjugate of the stable recurrent configurations in the single-toppling sandpile $\mathcal{A}_0$, coincides with the set of recurrent configurations of the sandpile in which $\mathcal{L}$ contains all the subsets of $[n]$. In this case, there are no stable transient configurations, and the condition for z being stable w.r.t. any toppling I, i.e., $z|_I \not\geq \Delta_I$ for all non-empty $I \subseteq [n]$, is related, by conjugation, to the condition of not having FSC's in $\mathcal{A}_0$, i.e., $(\iota z)|_I \not\geq f_{\max}(I)$ (because $f_{\max}(I) = (\bar{z} - \vec{\Delta}_I)|_I$, and, applying the definition of ι, we should recall that the multitoppling sandpile is formulated in the positive-cone gauge).

Now consider an 'anomalous' relaxation process, ρ_I, which may perform a multi-toppling rule I only at the first step, if possible, and then perform a single-toppling relaxation with $\mathcal{R}_0$. Such a process is unambiguous, but different processes may not commute, i.e., $\rho_I(\rho_J(z)) \neq \rho_J(\rho_I(z))$ in general.

Nonetheless, take a whatever semi-infinite sequence $(I_1, I_2, I_3, \ldots)$ of elements in $\mathcal{L} \setminus \mathcal{L}_0$, such that all elements in $\mathcal{L} \setminus \mathcal{L}_0$, occur infinitely-many times. It is easy to see that, for all z, there exists a truncation time $t = t(z)$ such that, for all $s \geq t$, $\rho_{I_s} \rho_{I_{s-1}} \cdots \rho_{I_1}(z) = \mathcal{R}(z)$, and in particular $\rho_I \mathcal{R}(z) = \mathcal{R}(z)$ for all $I \in \mathcal{L}$.

The interest in these processes ρ_I is in the fact that their action is strongly related to the action of the idempotents $\Pi_i = a_i^\dagger a_i$. Before stating and proving this relation in precise terms, it is instructive to investigate first how this works in the case of the BTW model. One easily recognizes that $\Pi_i z = z$ if $z_i < 3$, or $z_i = 3$ and $z_j < 3$ for all the neighbors j of i. In the first case, no topplings or antitopplings are involved, while in the second case exactly one toppling and one antitoppling at i occur. Conversely, if $z_i = 3$ and $z_j = 3$ for some neighbor j, a_i causes an avalanche for which a valid sequence of topplings may start with $(i, j, \ldots)$, i.e., $a_i z = t_{i_k} \cdots t_{i_3} t_j t_i z$ for some $(i_3, \ldots, i_k)$. The *effect* of the two initial topplings is identical to the effect of a multitoppling at a pair $\{i, j\}$. And, crucially, also the *if condition* coincides with the one for a configuration to be unstable w.r.t. the multitoppling at $\{i, j\}$. Thus, a configuration $z \in \mathbb{Z}^n$ is left stable by the application of Π_i if and only if it is stable w.r.t. both the toppling i and the multitopplings $\{i, j\}$ for all j neighbors of i. This proves Theorem 3.5 in the case of the BTW sandpile, and characterizes $y(z, I)$ as the result of the relaxation of z, in the multitoppling sandpile for which $\mathcal{L} \smallsetminus \mathcal{L}_0 = \big\{\{i, j\}\big\}_{i \in I, |j-i|=1}$.

The proof in the general setting, that we present below, is completely analogous.

Proof of Theorem 3.5. One finds that $\Pi_i z = z$ if $z_i < \overline{z}_i$, or $z_i = \overline{z}_i$ and $z_j \leq \overline{z}_j + \Delta_{ij}$ for all $j \neq i$. Again, in the first case no topplings or antitopplings are involved, while in the second case exactly one toppling and one antitoppling at i occur.

Conversely, if the conditions above are violated, a_i causes an avalanche for which a valid sequence of topplings may start with $(i, j, \ldots)$, and the effect of the two initial topplings is identical to the effect of a multitoppling at a pair $\{i, j\}$.

The if condition for the multitoppling $\{i, j\}$ to occur is that $z_i > \overline{z}_i + \Delta_{ji}$ and $z_j > \overline{z}_j + \Delta_{ij}$. The condition for the avalanche to involve topplings at i and j is $z_i = \overline{z}_i$ and $z_j > \overline{z}_j + \Delta_{ij}$. These two sets of conditions are certainly simultaneously not satisfied if $\Delta_{ij} = 0$, thus we can restrict our attention to sites j such that $\Delta_{ij} < 0$. In this case, the two sets coincide if and only if $\Delta_{ji} = -1$. Thus, in order to make the sets coincide for all sites i, we need that Δ is symmetric, and all the non-zero off-diagonal entries are -1, as required in the theorem hypotheses.

The configuration $y(z, I)$ is the result of the relaxation of z, in the multitoppling sandpile for which $\mathcal{L} \smallsetminus \mathcal{L}_0 = \big\{\{i, j\}\big\}_{i \in I, \Delta_{ij}=-1}$. $\square$

3.4 Wild Orchids: A Markov Chain Dynamics Involving Both Sand Addition and Removal

In this section we discuss a dynamics involving the operators a_i and $a_i^\dagger$. To keep the visualization simple, all our examples are variations of the BTW model, on portions of the square lattice and with heights in the range $\{0, 1, 2, 3\}$.

The dynamics starts from the maximally filled configuration, $z_i = 3$ for all i, and acts with the idempotent combinations $\Pi_i = a_i^\dagger a_i$ at randomly-chosen sites.

As we know from Theorem 3.5, this dynamics is absorbent on a unique configuration, identified with the multitoppling relaxation of the initial configuration, when pairs of adjacent sites both with height 3 are unstable.

Interesting features emerge at short times, when the configuration takes the form of a "web of strings", satisfying a classification theorem and a collection of incidence rules, first explained in [12], that are presented in Chap. 5.

On the other extremum of the dynamics, at the fixed point we have configurations that we call *Wild Orchids* and show remarkable regularities, in the form of 'patches', that is, a local two-dimensional periodicity on portions of the domain, a phenomenol-

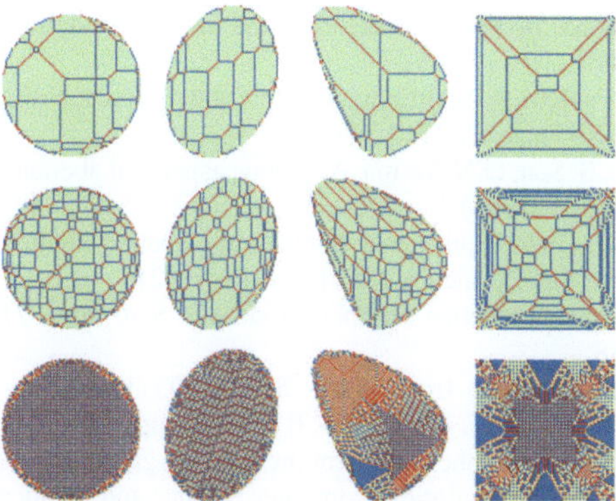

Fig. 3.1 Configurations obtained with the procedure described in Sect. 3.4. *Top*, *middle* and *bottom row* correspond to $t = 32$, $t = 1024$, and to the fixed point of the dynamics, the Wild Orchid. The columns show different domains. From *left* to *right*: a circle; an ellipse with axes rotated by arctan $1/3$ w.r.t. the cartesian axes, and ratio 2 between height and width; a typical algebraic curve of degree 3, more precisely $4x^2 + 4y^2 + 3xy + 2x^2y + x^3 = 7$ and a square. Color code is given in Fig. 5.1

ogy first investigated in [13], which will be studied in Chap. 5. When more patches are present, they follow an incidence rule first proven in [14]. Furthermore, being the final configuration stable for the dimer multitoppling rule, the sites i with height equal to 3 are isolated.

If the initial domain is an elliptic portion of the square lattice, a specially higher regularity emerges. Say that the linear dimension of the domain is of order L, and the slope of the symmetry axes is a "small" rational p/q (with both p and q of $\mathcal{O}(1)$ in L). Then, in the limit $L \to \infty$, we observe the emergence of a very simple structure of patches and strings: we have a unique patch, crossed by strings of a unique type, parallel to one of the two symmetry axes. This fact is in agreement with the general theory developed in [12–14], as the toppling vector at a coarsened level is a quadratic form in the coordinates x and y, that should vanish at the boundary of the domain, and the contour lines of quadratic forms are conics, i.e., plane algebraic curves of degree 2.

In Fig. 3.1 we present configurations obtained with the procedure described above, starting with the maximally-filled configuration, $z_i = 3$ for all i, on portions of the square lattice of various shapes: a disk, an ellipse, a smooth domain which is not a conic (it is an algebraic curve of degree 3), and a square. These figures show clearly the properties described above.

References

1. R. Karmakar, S.S. Manna, Particle-hole symmetry in a sandpile model. J. Stat. Mech: Theory Exp. **2005**, L01002 (2005)

2. D. Dhar, S.S. Manna, Inverse avalanches in the abelian sandpile model. Phys. Rev. E **49**, 2684–2687 (1994)
3. S. Caracciolo, G. Paoletti, A. Sportiello, Multiple and inverse topplings in the abelian sandpile model, preprint, (2011), arXiv:1112.3491.
4. D. Dhar, P. Ruelle, S. Sen, D.N. Verma, Algebraic aspects of abelian sandpile models. J. Phys. A: Math. Gen. **28**, 805 (1995)
5. S.N. Majumdar, D. Dhar, Equivalence between the abelian sandpile model and the $q \to 0$ limit of the potts model. Physica A **185**, 129–145 (1992)
6. E.R. Speer, Asymmetric abeiian sandpile models. J. Stat. Phys. **71**, 61–74 (1993). doi:10. 1007/BF01048088
7. M. Creutz, Abelian sandpiles. Comp. Phys. **5**, 198–203 (1991)
8. M. Creutz, Abelian sandpiles, Nucl. Phys. B (Proc. Suppl.), 20 (1991), pp. 758–761.
9. Y. Le-Borgne, D. Rossin, On the identity of the sandpile group, Discrete Mathematics. LaCIM 2000 Conference on Combinatorics, Computer Science and Applications, 256, pp. 775–790 2002
10. R. Cori, D. Rossin, On the sandpile group of dual graphs. Eur. J. Comb. **21**, 447–459 (2000)
11. J. M. Howie, An introduction to semigroup theory, Academic Press,London 1976.
12. S. Caracciolo, G. Paoletti, A. Sportiello, Conservation laws for strings in the abelian sandpile model EPL (Europhysics Letters) **90 (2010), p. 60003, arXiv:1002.3974v1**
13. S. Ostojic, Patterns formed by addition of grains to only one site of an abelian sandpile. Physica A **318**, 187–199 (2003)
14. D. Dhar, T. Sadhu, S. Chandra, Pattern formation in growing sandpiles. EPL (Europhys. Lett.) **85**, 48002 (2009)

Chapter 4
Identity Characterization

In this chapter we will present the derivation of an Explicit formula for the *Identity configuration* of the Abelian Sandpile Model in a particular directed lattice, the *Pseudo-Manhattan lattice*, that is known in literature also under the name of *F-lattice* [1]. This is the first explicit characterization of an Identity configuration for the ASM.

The results presented in this chapter are published in [2].

4.1 Introduction

4.1.1 Identity and Patterns

The main reason of our study of the Identity configurations, that can be seen in Fig. 4.1, comes from the fact that in this configuration it is possible to find intriguing and interesting patterns that cover the different part of it, in some kind of triangular shapes that resemble to be scale invariant. Since the first studies of Creutz [3, 4] there has been much interest on this identity configuration, Creutz itself shows in a later paper [5] how many interesting properties can be found in the ASM, and bring as example the calculation of the identity configuration.

What we want to stress is the emergence, in the identity, of patches covered with periodic patterns, which are scale invariant. Sometimes this patches are crossed by unidimensional defects; these defects and the patches themselves will be treated in detail in Chap. 5. This kind of patterns which are shown in the identity are similar to the ones showed when adding particles on the sandpile on a single site, as has been studied by Dhar et al. in [1]. These features are some of the reason that pushed us toward the study of this particular configuration of the sandpile.

G. Paoletti, *Deterministic Abelian Sandpile Models and Patterns*,
Springer Theses, DOI: 10.1007/978-3-319-01204-9_4,
© Springer International Publishing Switzerland 2014

Fig. 4.1 Identities on square geometry $L \times L$ on the BTW for different sizes, from *left* to *right* and *up* to *down*, L equal to 32, 64, 128 and 256. Different colors correspond to different height with the key of Fig. 5.1

4.1.2 ASM: Some Mathematics

As already discussed in Chap. 2 the ASM has a deep mathematical structure. In particular thanks to the Markov dynamics, the subspace of *Recurrent* configurations **R** emerges and here a mathematical group structure can be defined.

Indeed the set **R** has an underlying abelian structure, for which a presentation is explicitly constructed in terms of the matrix Δ, through the (heavy) study of its Smith normal form. Shortcuts of the construction of the presentation and more explicit analytical results have been achieved in the special case of a rectangular $L_x \times L_y$ portion of the square lattice, and still stronger results are obtained for the case of $L_x = L_y$ [6].

Call $\tilde{a}_i$ the operator which adds a particle at site i to a configuration C, and a_i the formal operator which applies $\tilde{a}_i$, followed by a sequence of topplings which makes the configuration stable. Remarkably, the final configuration $a_i C$ is independent from the sequence of topplings, and also, applying two operators, the two configurations $a_j a_i C$ and $a_i a_j C$ coincide, so that at a formal level a_i and a_j do commute.

More precisely, if a_i acting on C consists of the fall of a particle in i, $\tilde{a}_i$, and the sequence of topplings $t_{i_1}, \ldots, t_{i_k}$ on sites $i_1, \ldots, i_k$, the univocal definition of a_i and the commutation of a_i and a_j follow from the two facts:

$$z_j \geq \bar{z}_j : \qquad t_j \tilde{a}_i C = \tilde{a}_i t_j C ; \tag{4.1}$$

$$z_i \geq \bar{z}_i \text{ and } z_j \geq \bar{z}_j : \qquad t_i t_j C = t_j t_i C . \tag{4.2}$$

Another consequence is that, instead of doing all the topplings immediately, we can postpone some of them after the following $\tilde{a}$'s, and still get the same result. Similar manipulations show that the relation

$$a_i^{\Delta_{ii}} = \prod_{j \neq i} a_j^{-\Delta_{ij}} \tag{4.3}$$

holds when applied to an arbitrary configuration.

These facts lead to the definition of an abelian semi-group operation between two configurations, as the sum of the local height variables z_i, followed by a relaxation process [3, 4, 7]:

$$C \oplus C' = \left(\prod_i a_i^{z_i} \right) C' = \left(\prod_i a_i^{z'_i} \right) C . \tag{4.4}$$

For a configuration C, we define multiplication by a positive integer:

$$k C = \underbrace{C \oplus \cdots \oplus C}_{k} . \tag{4.5}$$

The set $\mathbf{R}$ of recurrent configurations is special, as each operator a_i has an inverse in this set, so the operation above, restricted $\mathbf{R}$, is raised to a group operation, the structure of the group of recurrent configuration of a graph has been studied in [8]. According to the Fundamental Theorem of Finite Abelian Groups, any such group must be a "discrete thorus" $T = \mathbb{Z}_{d_1} \times \mathbb{Z}_{d_2} \times \cdots \times \mathbb{Z}_{d_g}$, for some integers $d_1 \geq d_2 \geq \cdots \geq d_g$, and such that $d_{\alpha+1}$ divides d_α for each $\alpha = 1, \ldots, g - 1$. The values d_α, called *elementary divisors* of Δ, and a set of generators e_α with the proper periodicities, can be constructed through the Smith Normal form decomposition [6]. The composition of whatever $C = \{z_i\}$ with the set R acts then as a translation on T. A further consequence is that, for any recurrent configuration C, the inverse configuration $(-C)$ is defined, so that $k C$ is defined for $k \in \mathbb{Z}$.

Consider the product over sites i of Eq. (4.3)

$$\prod_i a_i^{\Delta_{ii}} = \prod_i \prod_{j \neq i} a_j^{-\Delta_{ij}} = \prod_i a_i^{-\sum_{j \neq i} \Delta_{ji}} \tag{4.6}$$

On the set R, the inverses of the formal operators a_i are defined, so that we can simplify common factors in (4.6), recognize the expression for b^+, and get

$$\prod_i a_i^{b_i^+} = I \tag{4.7}$$

so that $\prod_i a_i^{b_i^+} C = C$ is a necessary condition for C to be recurrent (but it is also sufficient, as no transient configuration is found twice in the same realization of the Markov chain), and goes under the name of *identity test*.

This condition turns into an equivalent and computationally-cheaper procedure, called *burning test* see Sect. 2.4.2, of which a side product, in case of positive answer, is a spanning arborescence rooted on the vertices of the border. So, the burning test provides us a bijection between the two ensembles, of recurrent sandpile configurations and rooted arborescences with roots on the border. This is in agreement with Kirchhoff Matrix-Tree theorem, which states that the number of such arborescences is given by det Δ, while the number of recurrent configurations is known to be det Δ as the first step of the procedure which determines the elementary divisors d_α [9, 10].

If, for the graph identified by Δ, a planar embedding exists, with all sites i such that $b_i^+ > 0$ on the most external face, then the planar dual of a rooted arborescence coming from the burning test is a spanning tree on the planar-dual graph.

This is clear for the undirected case $\Delta = \Delta^T$, and needs no more words. If the graph is directed, the arborescence is directed in the natural way, while the dual tree is constrained in some complicated way (some combinations of edge-occupations are forced to fixed values). However, the graphical picture simplifies if, on the planar embedding, in-coming and out-coming edges are cyclically alternating (and all plaquettes have consistent clockwise or counter-clockwise perimeters), and all sites with either b^+ or b^- positive are on the most external face, with in- and out-going arrows cyclically alternating (cfr. for example Fig. 4.2). We will call a directed graph of this kind a *planar alternating directed graph*.

4.2 Identities

Given the algebraic relation (4.7), and the semi-group operation (4.4), one could define the *frame* configuration Id_f as the one with $z_i = b_i^+$ for all i, and realize that it acts as an identity on recurrent configurations, $Id_f \oplus C = C$ if $C \in \mathbf{R}$. Conversely,

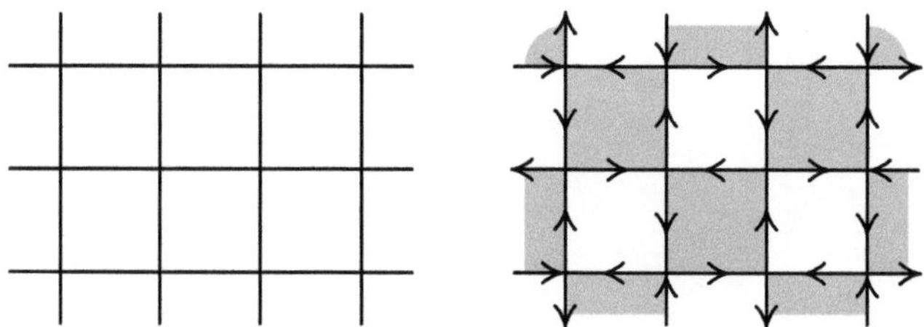

Fig. 4.2 *Left* a portion of the square lattice. *Right* a portion of the directed square lattice we considered in this work, with in- and out-edges alternated cyclically, and *white* and *gray* faces with *arrows* oriented clockwise and counter-clockwise

in general it does not leave unchanged a transient configuration, and in particular, as, for any relevant extensive graph, Id_f is itself transient, we have that $Id_f \oplus Id_f$, and $Id_f \oplus Id_f \oplus Id_f$, and so on, are all different, up to some number of repetitions k at which the configuration is sufficiently filled up with particles to be recurrent. We call Id_r this configuration, and $k(\Delta)$ the minimum number of repetitions of Id_f required in the ASM identified by Δ (we name it the *filling number* of Δ). The configuration Id_r is the identity in the abelian group $\mathbb{Z}_{d_1} \times \cdots \times \mathbb{Z}_{d_g}$ described above, and together with a set of generators e_α, completely identifies in a constructive way the group structure of the statistical ensemble. The relevance of this configuration has been stressed first by M. Creutz [3–5, 7], so that we shall call it *Creutz identity*. Some exact results on the decomposition of the identity in different parts have been achieved by Le Borgne and Rossin in [11]. See Figs. 4.3 and 4.1 for some examples.

Unfortunately, despite many efforts, it has not been possible to give a closed-formula recipe for this identity state on given large lattices, not even in the case of a $L \times L$ square, and the direct numerical investigation of these configurations at various sizes has produced peculiar puzzling pictures [12].

In a large-side limit, we have the formation of curvilinear triangular regions of extensive size (of order L), filled with regular patterns, and occasionally crossed by straight "defect lines", of widths of order 1, which, furthermore, occasionally meet at Y-shaped "scattering points", satisfying peculiar conservation laws as in [13] and in Chap. 5.

Similar features emerge also for the filling numbers, e.g. on the square lattice of size L, the index $k_L \equiv k(\Delta^{(L)})$ is not badly fitted, for even L, by a parabola $k_L \simeq L^2/6 + o(L^2)$, but showing fluctuations due to unpredictable number-theoretical properties of L. The challenging sequence of these numbers, for L up to 64, is given in Table 4.1. It should be noted that, conversely, odd sizes $2L + 1$ are related to $2L$ through a property proven in [6, sect. 4.7].

The determination of Id_r for a given graph is a procedure polynomial in the size of the graph. E.g. one could prove that $k(\Delta)$ is sub-exponential, and that relaxing $C \oplus C'$ for C and C' both stable is polynomial, then one can produce the powers $2^s Id_f$ recursively in s, up to get twice the same configuration. Better algorithms exist however, see for example [3, 4, 7].

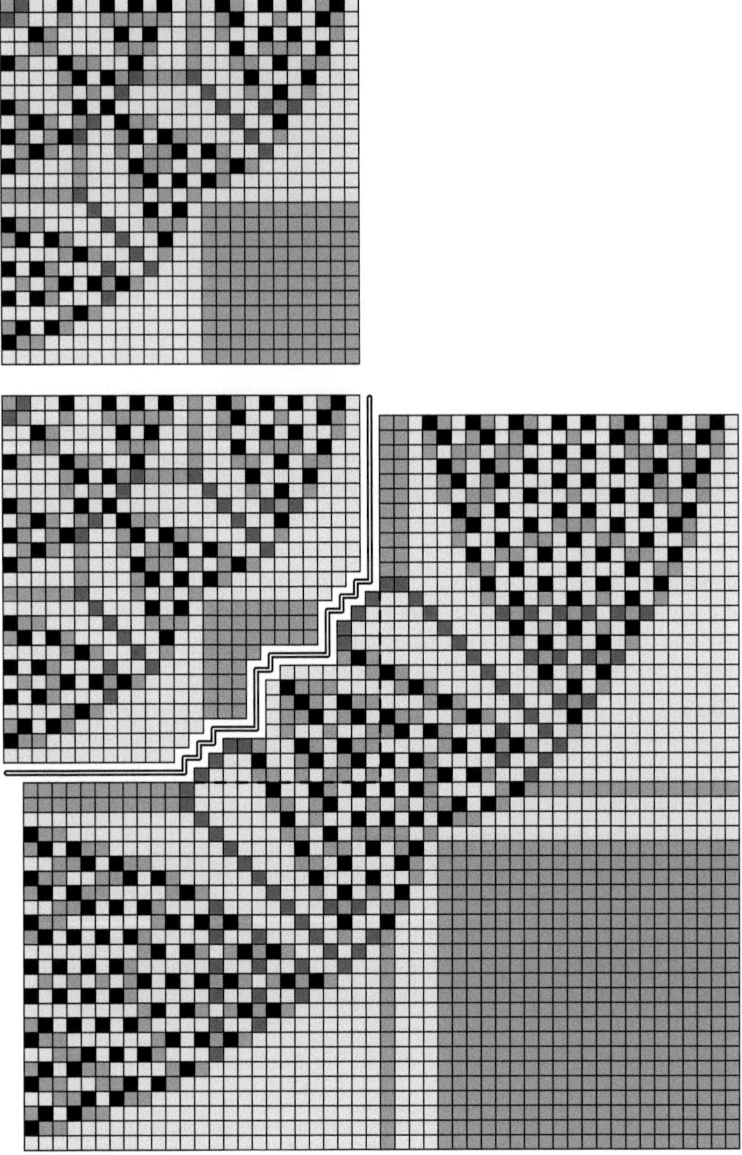

Fig. 4.3 The *top-left* corners of the identities $Id_r^{(L)}$ for $L = 50$ and $L = 100$ in the BTW model (the other quadrants are related by symmetry). Heights from 0 to 3 correspond to *gray tones* from *dark* to *light*. The smaller-size identity is partially reproduced at the corner of the larger one, in a fashion which resembles the results of Theorem 4.6

Table 4.1 Values of k_L for the BTW Abelian Sandpile on square geometries of even size, for $L = 2, \ldots, 64$

L	2	4	6	8	10	12	14	16	18	20	22	$\triangleright$
k_L	1	4	7	13	19	27	35	46	58	71	87	$\blacktriangleright$
$\triangleright$	24	26	28	30	32	34	36	38	40	42	44	$\triangleright$
$\blacktriangleright$	103	119	138	156	180	198	226	248	276	305	334	$\blacktriangleright$
$\triangleright$	46	48	50	52	54	56	58	60	62	64		
$\blacktriangleright$	367	397	430	464	499	538	572	615	653	699		

Still, one would like to have a better understanding of these identity configurations. This motivates to the study of the ASM on different regular graphs, such that they resemble as much as possible the original model, but simplify the problem in some regards so that the family (over L) of resulting identities can be understood theoretically.

We explore some possible different regular graphs, where is defined an ASM, and we choose the best candidates among them following a few principles: (1) we wish to simplify the problem in some structural way; (2) we want to preserve the property of the $L \times L$ lattice, discussed in [6, sect. 4.4] that allows to reconstruct the elementary divisors of Δ through a suitable matrix $n_{yy'}$, which appears when producing a presentation of the group, in terms of $\mathcal{O}(L)$ generators a_i of the whole set L^2; (3) we want to preserve planarity, and the interpretation of dual spanning subgraphs as spanning trees on the dual lattice, i.e. we want to use either planar undirected graphs, or planar alternating directed graphs (according to our definition above).

4.3 A Prolog: Cylindric Geometries

A first possible simplification comes from working on a cylindrical geometry. We call respectively *periodic*, *open* and *closed* the three natural conditions on the boundaries of a $L_x \times L_y$ rectangle. For example, in the BTW Model, for site $(i, 1)$, in the three cases of periodic, open and closed boundary conditions on the bottom horizontal side we would have the toppling rules

$$
\begin{aligned}
\text{periodic}: \quad & z_{i,1} \to z_{i,1} - 4; \quad && z_{i\pm1,1} \to z_{i\pm1,1} + 1; \\
& z_{i,2} \to z_{i,2} + 1; \quad && z_{i,L_y} \to z_{i,L_y} + 1; \\
\text{open}: \quad & z_{i,1} \to z_{i,1} - 4; \quad && z_{i\pm1,1} \to z_{i\pm1,1} + 1; \\
& z_{i,2} \to z_{i,2} + 1; \quad && \\
\text{closed}: \quad & z_{i,1} \to z_{i,1} - 3; \quad && z_{i\pm1,1} \to z_{i\pm1,1} + 1; \\
& z_{i,2} \to z_{i,2} + 1; \quad &&
\end{aligned}
$$

The external face, with $b_i^{\pm} \neq 0$, is in correspondence of open boundaries, so, as we want a single external face, if we take periodic boundary conditions in one direction (say, along x), the only possible choice in this framework is to take closed and open conditions on the two sides in the y direction.

Our notation is that $\bar{z}_i$ takes the same value everywhere: on open boundaries, b_i^- and b_i^+ are determined accordingly, while on closed boundaries either we add some extra "loop" edges, or we take $\bar{z}_i - \Delta_{ii} > 0$.

The cylindric geometry has all the non-trivial features of the original ASM for what concerns group structures, polynomial bound on the relaxation time in the group action, connection with spanning trees, and so on (even the finite-size corrections to the continuum-limit CFT are not more severe on a $L_x \times L_y$ cylinder than on a $L_x \times L_y$ open-boundary rectangle), *but*, for what concerns the determination of $Id_{\rm r}$ through relaxation of $kId_{\rm f}$, the system behaves as a quasi-unidimensional one, and $Id_{\rm r}$ is in general trivially determined.

Furthermore, in many cases $Id_{\rm r}$ is just the maximally-filled configuration $C_{\max}$. This fact is easily proven, either by checking that $Id_{\rm r} \oplus Id_{\rm f} = Id_{\rm r}$, which is easy in the quasi-1-dimensional formulation (or in other words, by exploiting the translation symmetry in one direction), or with a simple burning-test argument, in the cases in which on each site there is a single incoming edge from sites nearer to the border. Fig. 4.4 shows a few examples.

Such a peculiar property has a desiderable consequence on the issue of inversion of a recurrent configuration. Indeed, if C and C' are in $\mathbf{R}$, we have that

$$(-C) \oplus C' = Id_{\rm r} \oplus (-C) \oplus C' = (Id_{\rm r} - C) \oplus C' \tag{4.8}$$

where, if $C = \{z_i\}$, $(-C) = \{\tilde{z}_i\}$ is the inverse configuration we seek, but $Id_{\rm r} - C$ is simply the configuration with $\tilde{z}_i = \bar{z}_i - z_i - 1$, and has all non-negative heights.

4.4 Pseudo-Manhattan Lattice

From this section on, we will concentrate on square geometries, on the square lattice with edges having a given orientation, and all vertices having in- and out-degree equal to 2.

For this reason, we can work with $\bar{z}_i = 2$, so that $S = \{0, 1\}^V$ instead of $\{0, 1, 2, 3\}^V$ (a kind of simplification, as from "CMYK" colour printing to "black and white"). In this step we lose in general a bit of symmetry: e.g. on a square of size $2L$ we have still the four rotations, but we lose the reflections, which are arrow-reversing, while on a square of size $2L + 1$ we lose rotations of an angle $\pi/2$, and only have rotation of π.

Square lattices with oriented edges have already been considered in Statistical Mechanics, especially in the variant called "Manhattan Lattice" (i.e. with horizontal edges oriented east- and west-bound alternately on consecutive rows, and coherently

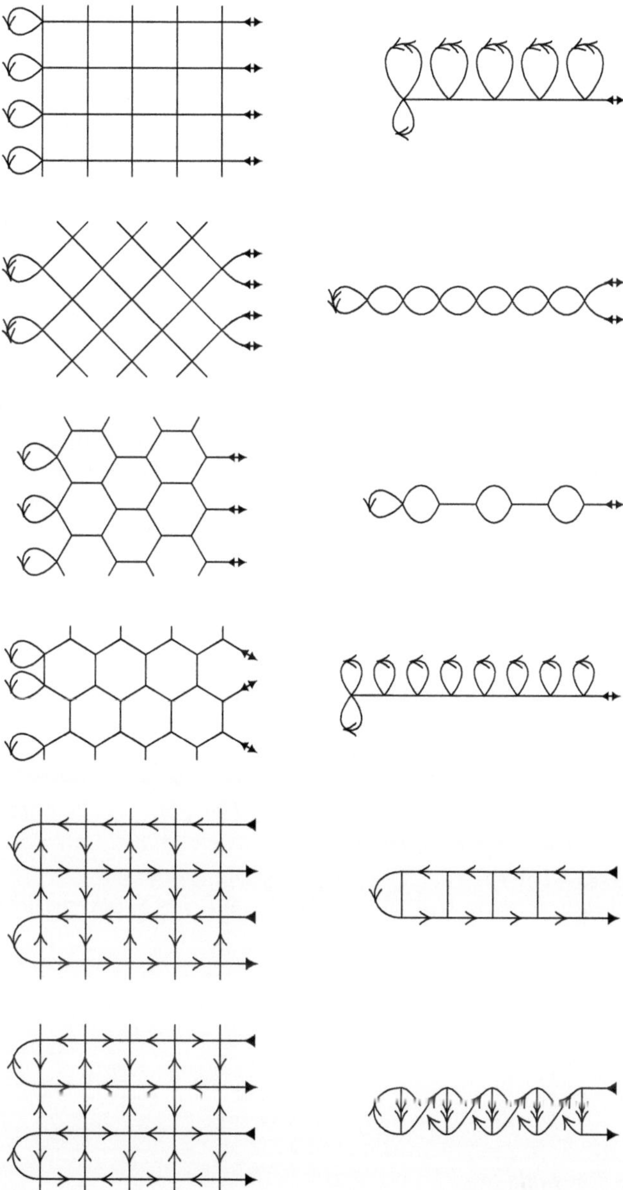

Fig. 4.4 Reduction to a quasi-1-dimensional system for the ASM on cylindric geometry, on a few examples all having Id_r coinciding with C_{max}. From *top* to *bottom*: a portion of the square lattice, in the two orientations; of the hexagonal lattice, in the two orientations, of Manhattan and pseudo-Manhattan lattices. Plain edges correspond to $\Delta_{ij} = \Delta_{ji} = -1$, while a directed edge (from i to j) with k *arrows* correspond to $\Delta_{ij} = -k$. In-(out-)coming *bold arrows* correspond to b^- (b^+) equal to 1, while the *lozenge-shaped double-arrows* correspond to $b^- = b^+ = 1$. A loop with k *arrows* on i corresponds to $\bar{z}_i - \Delta_{ii} = k$, otherwise $\bar{z}_i = \Delta_{ii}$

within a row, and similarly for vertical edges), cfr. for example [14]. However this lattice in two dimensions is not a planar alternating directed graph, and the results for the related ASM model will be discussed only briefly in last section.

We start by analysing a less common variant, which is better behaving for what concerns the ASM model, and which we call *pseudo-Manhattan lattice* (it appears, for example, in the totally unrelated paper [15]). In this case, the horizontal edges are oriented east- and west-bound alternately in both directions (i.e. in a chequer design), and similarly for vertical edges. As a result, all plaquettes have cyclically oriented edges. A small portion of this lattice is shown in Fig. 4.2, where it is depicted indeed as the prototype planar alternating directed graph. Conventionally, in all our examples (unless otherwise specified) we fix the orientations at the top-right corner to be as in the top-right corner of Fig. 4.2.

Quite recently, in [1, 16–18] both Manhattan and pseudo-Manhattan lattices have been considered as an interesting variant of the ASM model (the latter under the name of *F-lattice*), with motivations analogous to ours. This corroborates our claim that these variants are natural simplifications of certain features in the original BTW model.

The main feature of the Creutz identities, on square portions of the pseudo-Manhattan lattice with even side, is self-similarity for sides best approximating the ratio $1/3$, up to a trivial part, as illustrated in Fig. 4.5.

The precise statement is in the following Theorem 4.6. Call $Id_r^{(L)}$ the set of heights in the Creutz identity for the square of side $2L$, encoded as a $L \times L$ matrix for one of the four portions related by rotation symmetry. Say that index $(1, 1)$ is at a corner of the lattice, and index (L, L) is in the middle. We have

Theorem 4.1 *Say $L = 3\ell + s$, with $s = 1, 2, 3$. Then, $Id_r^{(L)}$ is determined from $Id_r^{(\ell)}$ and a closed formula, and thus, recursively, by a deterministic telescopic procedure in at most $\lfloor \log_3 L \rfloor$ steps.*

For $s = 1$ or 3, we have $\left(Id_r^{(L)}\right)_{ij} = \left(Id_r^{(\ell)}\right)_{ij}$ if $i, j \leq \ell$, otherwise $\left(Id_r^{(L)}\right)_{ij} = 0$ iff,

for $i + j + s$ even,

$$
\begin{align}
i < \ell, & \qquad |L - \ell - j| - i > 0; \tag{4.9a} \\
j > \ell, & \qquad |2\ell + 1 - i| + j < 2\ell + s; \tag{4.9b} \\
i \leq 2\ell, & \qquad j = L - \ell; \tag{4.9c}
\end{align}
$$

and, for $i + j + s$ odd,

$$
\begin{align}
i > \ell, & \qquad |L - \ell - i| - j > 0; \tag{4.9d} \\
i > \ell - 1, & \qquad |2\ell + s - j| + i < 2\ell; \tag{4.9e}
\end{align}
$$

If $s = 2$ the same holds with i and j transposed in $Id_r^{(L)}$ and $Id_r^{(\ell)}$ (but not in (4.9a–e)). Furthermore, $k_L = \frac{L(L+1)}{2}$.

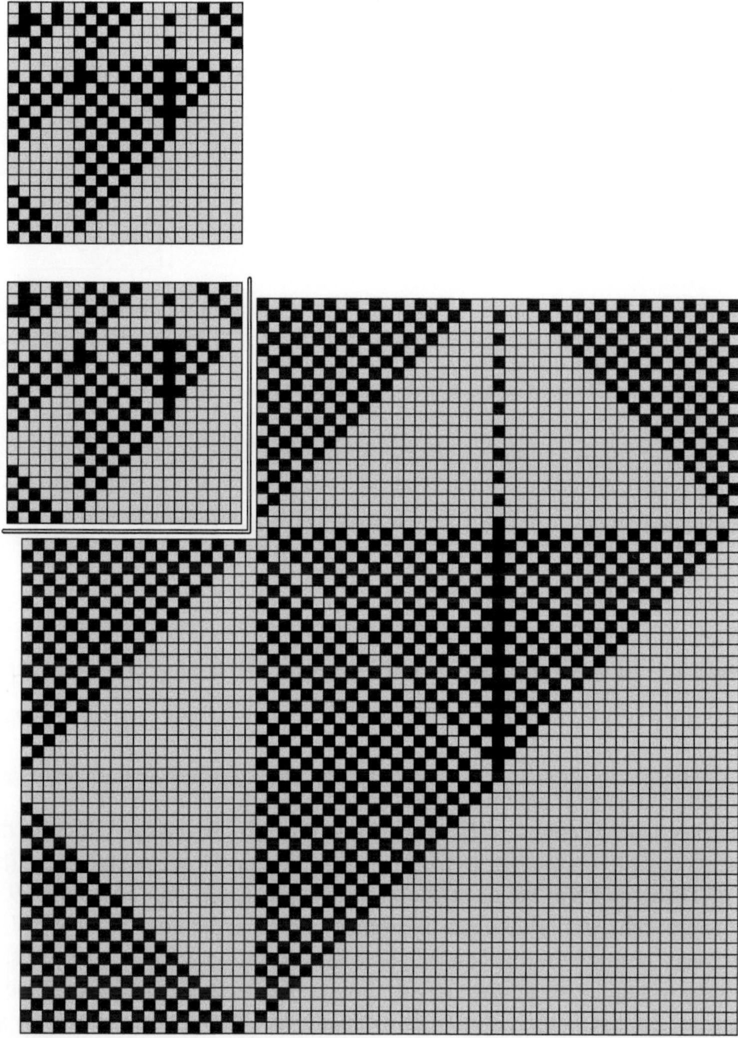

Fig. 4.5 The *top-right* corners of the identities $Id_{\mathrm{r}}^{(L)}$ for $L = 21$ and $L = 64$ (remark $64 = 3 \cdot 21 + 1$). The smaller-size one is exactly reproduced at the corner of the larger one, while the rest of the latter has an evident regular structure, according to Theorem 4.6

The statement of Eq. (4.9–e) is graphically represented in Fig. 4.6.

The understanding of the Creutz identity on square portions of the pseudo-Manhattan lattice is completed by the following theorem, relating the identity at side $2L + 1$ to the one at side $2L$. We encoded the identity at even sides in a $L \times L$ matrix $Id_{\mathrm{r}}^{(L)}$, exploiting the rotation symmetry, such that the extended $2L \times 2L$ matrix has the property

$$(Id_{\mathrm{r}}^{(L)})_{i,j} = (Id_{\mathrm{r}}^{(L)})_{2L+1-j,i} \, . \tag{4.10}$$

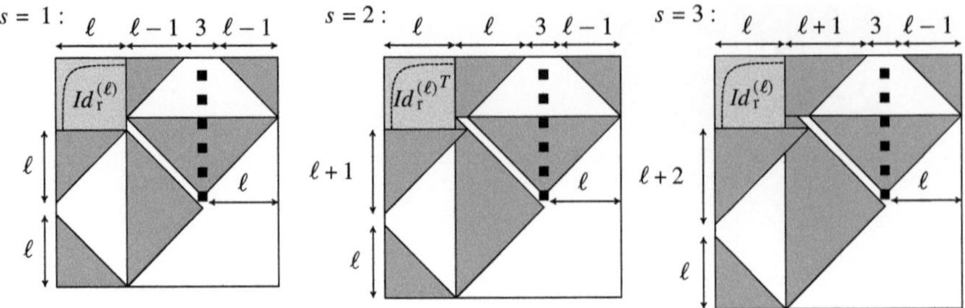

Fig. 4.6 The non-recursive part of the identities $Id_{\mathrm{r}}^{(L)}$ (*top-right* corners) for $L = 3\ell + s$, and $s \in \{1, 2, 3\}$, illustrating the set described by (4.9a–e). For $s = 2$, we show the transposed of Id_{r}. *Black* and *white* stand respectively for $z_i = 0$ and 1; *gray regiorns* corresponds to chequered parts, starting with *white* on the cells cutted by $\pi/4$-inclination lines

We can similarly encode the identity at odd sides in a structure of almost $1/4$ of the volume, namely a $(L + 1) \times (L + 1)$ matrix $\hat{Id}_{\mathrm{r}}^{(L)}$, using the fact that

$$(\hat{Id}_{\mathrm{r}}^{(L)})_{i,j} = (\hat{Id}_{\mathrm{r}}^{(L)})_{2L+2-i,j} = (\hat{Id}_{\mathrm{r}}^{(L)})_{i,2L+2-j} \,. \tag{4.11}$$

Theorem 4.2 *We have* $(\hat{Id}_{\mathrm{r}}^{(L)})_{i,j} = (Id_{\mathrm{r}}^{(L)})_{i,j}$ *if* $i, j \leq L$, $(\hat{Id}_{\mathrm{r}}^{(L)})_{L+1,j} = 0$ *if* $L - j$ *is odd and 1 if it is even, and, for* $i \leq L$, $(\hat{Id}_{\mathrm{r}}^{(L)})_{i,L+1} = 0$ *if* $L - i$ *is even and 1 if it is odd.*

The statement of this theorem is illustrated in Fig. 4.7.

The proof of Theorem 4.7, given Theorem 4.6, is easily achieved through arguments completely analogous to the ones of [6, sect. 7]. We postpone the (harder) proof of Theorem 4.6 to the discussion of the equivalence with Theorem 4.9 below.

There are some similarities and differences with the identities in the BTW model. In that case, besides the evident height-2 square in the middle, there arise some curvilinear triangolar structures, mostly homogeneous at height 3, but some others "texturized", so that there exists a variety of patterns which appear extensively at large sizes (in the metaphor of CMYK offset printing, like the way in which composite colours are produced!). A first attempt of classification of these structures appears in [19]; some progresses were made by [13] and a wider discussion on the classification, coming to a complete list, is given in Chap. 5. In our case, we only have "black and white", but, as four zeroes in a square are a forbidden configuration (as well as many others too dense with zeroes), we can not have extesive regions of zeroes (black, in our drawings of Fig. 4.5). A big square in the middle is still there, rotated of $45°$, while triangoloids are replaced by exact "45–45–90" right triangles in a "texturized gray" coming from a chequered pattern. Indeed, to our knowledge, such a regular structure as in theorem 4.6 was not deducible *a priori* (in particular, not before the publication of [1]), and our initial motivation was to study the emergence of patterns in a 2-colour case.

Fig. 4.7 The recurrent identity on the pseudo-Manhattan lattice of side 50 (*up*) and 51 (*down*), an example of how the identity on side $2L + 1$ is trivially deduced from the one on side $2L$

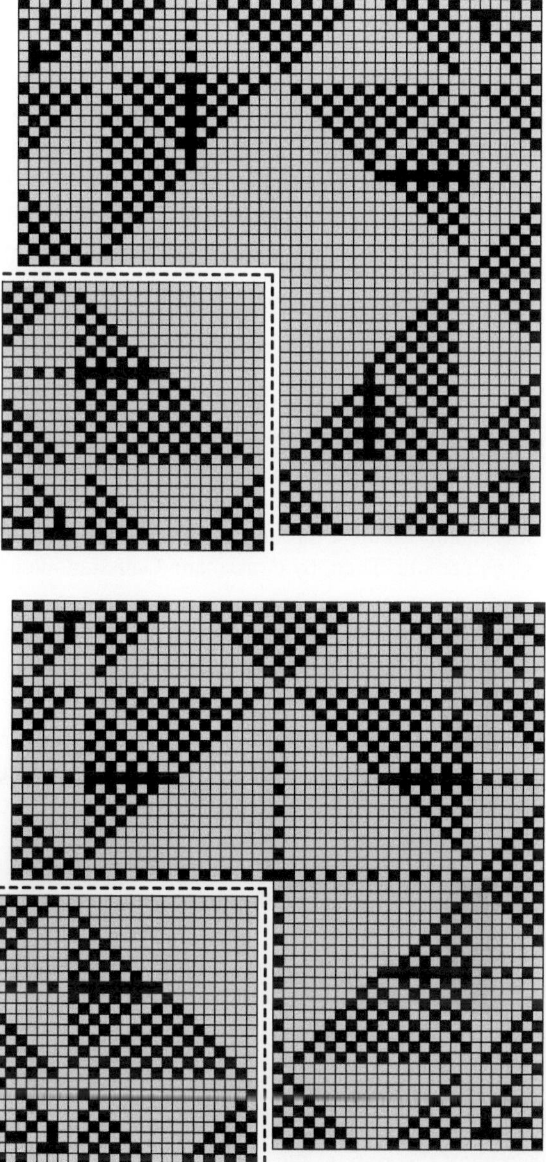

The statement of Theorem 4.6 would suggest to look for similar features also in the BTW model. It turns out that, while in the directed case the $\lfloor (L - 1)/3 \rfloor$ size is *fully* contained in the corner of the L size, in the BTW the $\lfloor L/2 \rfloor$ size is *partially* contained in the corner of the L size, in an empirical way which strongly fluctuates with L, but is in most cases more than 50 % (cfr. Fig. 4.3 for an example).

The heuristics in the case of the Manhattan lattice, under various boundary prescriptions, are somewhat intermediate: the continuum-limit configuration exists and coincides with the one for the pseudo-Manhattan, but the $\lfloor (L-1)/3 \rfloor$ corners are only partially reproduced by the size-ℓ identities.

How could one prove Theorem 4.6? An elegant algebraic property of the identity is that it is the only recurrent configuration for which all the "charges" are zero (see [6], Eqs. (3.3), (3.4)). More precisely, and in a slightly different notation, given a whatever ordering of the sites, and any site i, and calling $A_{i,j}$ the minor (i,j) of a matrix A, we have

$$Q_j(C) := \sum_i z_i (-1)^{i+j} \det \Delta_{i,j} \tag{4.12}$$

and $Q_j(Id_\mathrm{r}) = q_j \det \Delta$ with $q_j \in \mathbb{Z}$, with Id_r being the only stable recurrent configuration with this property.

A possible proof direction (that we do *not* follow in this chapter) could have been as follows. An algebraic restatement of the expression for the charges is achieved in Grassmann calculus, through the introduction of a pair of anticommuting variables $\bar{\psi}_i$, ψ_i per site. Then, by Grassmann Gaussian integration, we have that

$$(-1)^{i+j} \det \Delta_{i,j} = \int \mathcal{D}(\psi, \bar{\psi}) \, \bar{\psi}_i \psi_j \, e^{\bar{\psi} \Delta \psi} \tag{4.13}$$

where here is a contribution $b_k^- \bar{\psi}_k \psi_k$ in the exponential for each site on the border, and a contribution $(\bar{\psi}_h - \bar{\psi}_k)\psi_k$ if a particle falls into h after a toppling in k, i.e. $\Delta_{hk} = -1$ (we are pedantic on this because the asymmetry of Δ could create confusion on who's who with $\bar{\psi}$ and ψ).

So the configurations in the same equivalence class of the identity are the only ones such that, for each i, the "expectation value"

$$\left\langle \left(\sum_j z_j \bar{\psi}_j \right) \psi_i \right\rangle := \frac{\int \mathcal{D}(\psi, \bar{\psi}) \, (\sum_j z_j \bar{\psi}_j) \psi_i \, e^{\bar{\psi} \Delta \psi}}{\int \mathcal{D}(\psi, \bar{\psi}) \, e^{\bar{\psi} \Delta \psi}} \tag{4.14}$$

is integer-valued.

Furthermore, expressions as in the right hand side of (4.13) are related, through Kirchhoff theorem, to the combinatorics of a collection of directed spanning trees, all rooted on the boundary, with the exception of a single tree which instead contains both i and j, and the path on the tree from i to j is directed consistently.

So, a possible approach by combinatorial bijections could be to prove that, for any j, there is a suitable correspondence among the forests as above, and a number q_j of copies of the original ensemble of rooted forests.

Such a task is easily performed, even for a generic (oriented) graph, for what concerns the frame identity Id_f, for which all charges q_j are 1. Unfortunately, for what concerns Id_r, and with an eye to the proof performed in the following section, it seems difficult to pursue this project at least in the case of the pseudo-Manhattan lattice on a square geometry. Indeed, if the relaxation of $k_L Id_f$ to Id_r requires t_j topplings on site j, it is easy to see that $q_j(Id_r) = k_L - t_j$, and from the explicit expressions for k_L and the values of the topplings (the latter are in the following Theorem 4.9) we see that the values of the charges q_j are integers of order L^2.

4.5 Proof of the Theorem

Here we perform the direct proof of Theorem 4.6. It is fully constructive, a bit technical, and maybe not specially illuminating for what concerns the algebraic aspects of the problem, but still, it makes the job.

Now, in order to better exploit the geometry of our square lattice, we label a site through a pair ij denoting its coordinates. We can introduce the matrix $T_{ij}^{(L)}$, which tells how many topplings site ij performed in the relaxation of $k_L Id_f$ into Id_r. Exploiting the rotational symmetry, we take it simply $L \times L$ instead of $2L \times 2L$, with $(i, j) = (1, 1)$ for the site at the corner, analogously to what we have done for $(Id_r^{(L)})_{ij}$. Of course, although we call $T^{(L)}$ and $Id_r^{(L)}$ "matrices", they have a single site-index, and are indeed vectors, for example, under the action of Δ.

Clearly, T is a restatement of Id_r, as

$$(Id_r^{(L)})_{ij} = k_L\, b_{ij}^+ - \Delta_{ij,i'j'}\, T_{i'j'}^{(L)} \tag{4.15}$$

so that Id_r is determined from T, but also vice-versa, as Δ is invertible. Actually, through the locality of Δ, one can avoid matrix inversion if one has some "boundary condition" information on T, and the exact expression for Id_r, e.g. if one knows T on two consecutive rows and two consecutive columns (and we have a guess of this kind, as we discuss in the following).

The constraint that $(Id_r)_{ij} \in \{0, 1\}$ gives that T is locally a parabola with small curvature. Moreover, in the regions corresponding to homogeneous portions of Id_r, T must correspond to a discretized parabola through easy formulas. The telescopic nature of $(Id_r^{(L)}, Id_r^{(\lfloor \frac{L-1}{3} \rfloor)}, \ldots)$ has its origin in an analogous statement for $(T^{(L)}, T^{(\lfloor \frac{L-1}{3} \rfloor)}, \ldots)$, and on the fact that k_L has a simple formula. As we will see, these facts are easier to prove.

We start by defining a variation of T which takes in account explicitly both the height-1 square in the middle of Id_r, and the spurious effects on the border. Define $\tilde{k}(L) = \lfloor (L-1)(L+2)/4 \rfloor$ (which is approximatively $k_L/2$), and introduce

$$\hat{T}_{ij}^{(L)} := T_{ij}^{(L)} - \tilde{k}(L)b_{ij}^+ - \binom{L-i+1}{2} - \binom{L-j+1}{2}. \tag{4.16}$$

A first theorem is that

Theorem 4.3

$$\hat{T}_{iL}^{(L)} = \hat{T}_{Lj}^{(L)} = 0; \tag{4.17a}$$

$$\hat{T}_{ij}^{(L)} = 0 \qquad for \; i + j > L + \ell; \tag{4.17b}$$

$$\hat{T}_{ij}^{(L)} \leq 0 \qquad for \; all \; i, j. \tag{4.17c}$$

This implicitly restates the claim about the middle square of height 1 in Id_{r}, and is in accord with the upper bound on the curvature of T given by $(Id_{\mathrm{r}})_{ij} \leq 1$. With abuse of notations, we denote by $\hat{T}_{ij}^{(\ell)}$ also the $L \times L$ matrix corresponding to $\hat{T}_{ij}^{(\ell)}$ in the $\ell \times \ell$ corner with $i, j \leq \ell$, and zero elsewhere. Then the rephrasing of the full Theorem 4.6 is

Theorem 4.4 *If* $M_{ij}^{(L)} = -(\hat{T}_{L-i \; L-j}^{(L)} - \hat{T}_{L-i \; L-j}^{(\ell)})$ *for* $s = 1, 3$ *and the transpose of the latter for* $s = 2$, *defining* $\theta(n) = 1$ *for* $n > 0$ *and* 0 *otherwise, and the "quadratic + parity" function* $q(n)$ *on positive integers*

$$q(n) = \theta(n) \left(n^2 + 2n + \frac{1 - (-1)^n}{2} \right)$$

$$= \begin{cases} 0 & n \leq 0; \\ (n+1)^2/4 & n \; positive \; odd; \\ n(n+2)/4 & n \; positive \; even; \end{cases} \tag{4.18}$$

then $M^{(L)}$ *is a deterministic function* [1], *piecewise "quadratic + parity" on a finite number of triagular patches*

$$M_{ij}^{(L)} = \binom{j+1}{2} + \binom{i+1}{2} + \max\left(0, \left\lfloor \frac{\ell + 1 - |i - \ell| - |j - L - \ell|}{2} \right\rfloor\right)$$

$$+ \theta(L - \ell - i)\binom{j - L + \ell + 1}{2} + \theta(L - \ell - j)\binom{i - L + \ell + 1}{2}$$

$$- q(i - j - L - \ell - 1) - q(j - i - L - \ell - 1)$$

$$- q(j + i - L + \ell) - q(i + j - L - \ell - 1). \tag{4.19}$$

Clearly Theorem 4.8 is contained in Theorem 4.9, just by direct inspection of the summands in (4.19). The equivalence among Theorems 4.6 and 4.9 is achieved still by direct inspection of (4.19), with the help of some simple lemmas. First remark that the use of $-\hat{T}$ instead of T makes us work "in false colours", i.e. effectively interchanges z into $1 - z$ in Id_{r}. Then, defining $\nabla_x^2 f(i, j) := f(i+1, j) + f(i-1, j) - 2f(i, j)$, and analogously ∇_y^2 with ± 1 on j, we have that

---

[1] The formula here present is different from the one presented in [2], this one is the correction.

$$\nabla^2_{x,y}q(\pm i \pm j - a) = \begin{cases} 1 & \pm i \pm j - a \geq 0 \text{ and even}; \\ 0 & \text{otherwise}; \end{cases} \qquad (4.20)$$

and that $\nabla^2_x \max(0, j - a) = 1$ at $j = a$ only, while $\nabla^2_y \max(0, j - a) = 0$ always (this reproduces the ℓ extra zeroes out of the triangles depicted in Fig. 4.6).

So, the explicit checks above can lead to the conclusion that Eq. (4.15) holds at every L for the expressions for T and Id_r given, in a special form, i.e. subtracting the equation for size L with the one for size ℓ, but on the $L \times L$ system. The latter is then easily related to the one on the $\ell \times \ell$ system, at the light of Eq. 4.17a–c, which allows to state that there are no different contributions from having splitted the four quadrants of $\hat{T}^{(\ell)}$. Thus, Eq. 4.17a–c, guessing the exact expression for T_{ij} on the edges of each quadrant by mean of a simple formula, is crucial to the possibility of having the telescopic reconstruction procedure.

In other words, even without doing the tedious checks, we have seen how, by a series of manipulations corresponding to the subtraction of $\ell \times \ell$ corners to the $L \times L$ matrices, the proof is restricted to the analysis of the "deterministic" part, with $i > \ell$ or $j > \ell$. As, in this case, all the involved functions depend on L through its congruence modulo 2, or 3, or 4, it suffices to check numerically the theorems for 12 consecutive sizes in order to have that the theorem must hold for all sizes. We did the explicit check for sizes up to $L = 64$.

The only thing that we need in order to conclude that the conjectured expression Id_r corresponds to the identity is to prove that it is indeed a recurrent configuration. Again, we do that in two steps, in order to divide the behaviour on the self-similar $\ell \times \ell$ corner and on the deterministic part.

First, remark that in $Id_r^{(L)}$ the internal border of the quadrant (the sites ij with i or $j = L$) is burnt in the Burning Test even without exploiting the other mirror images. This is true because, at every L and on both borders, we have a substructure of the form

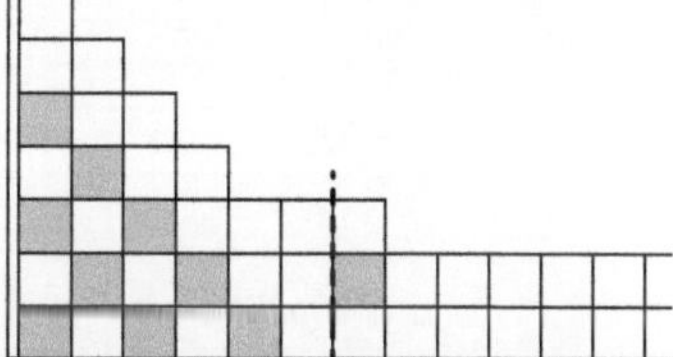

which is burnt through the following cascade (we put the burning times, and denote with arrows from x to y a toppling occurring in x which triggers the toppling in y)

Fig. 4.8 The recurrent identity on a portion of side 50 of the square lattice, with pseudo-Manhattan (*up*) and Manhattan (*down*) orientation

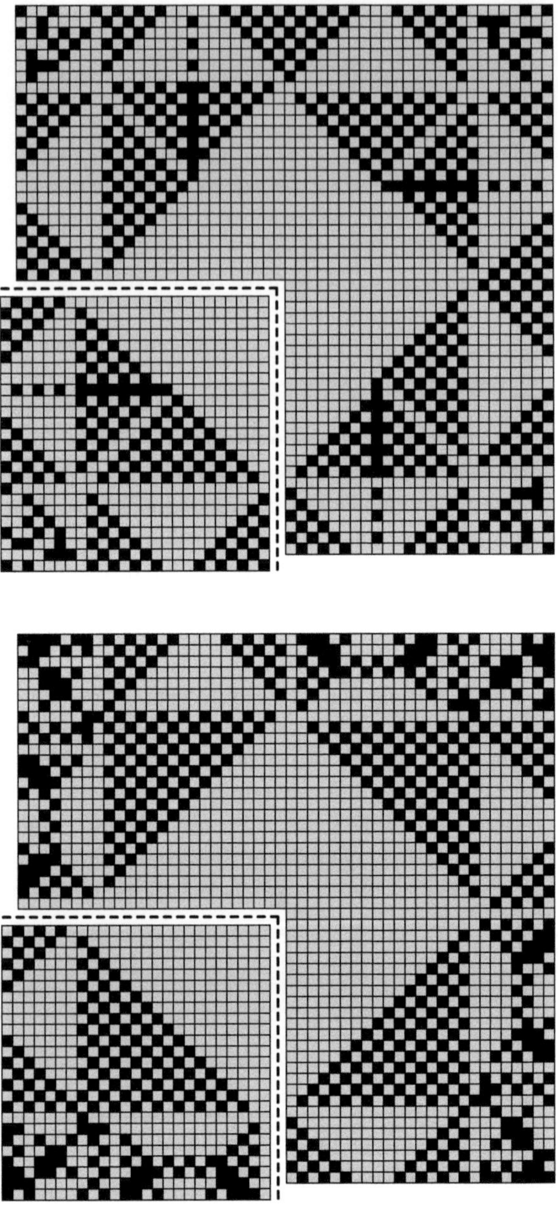

We choose to show just an example, instead of giving the explicit formula, but it should be clear from the pictures above that a general-L regular procedure exists.

So, Id_r satisfies the burning test if and only if the deterministic part of Id_r (the $L \times L$ square minus the $\ell \times \ell$ corner) satisfies the burning test with the border

Fig. 4.9 *Up* the limit Creutz Identity configuration on our Manhattan and pseudo-Manhattan lattices. *White* regions correspond to have height 1 almost everywhere. *Gray* regions correspond to have height 0 and 1 in a chequered pattern almost everywhere. *Down* the image of a quadrant under the map $z \to \ln z$

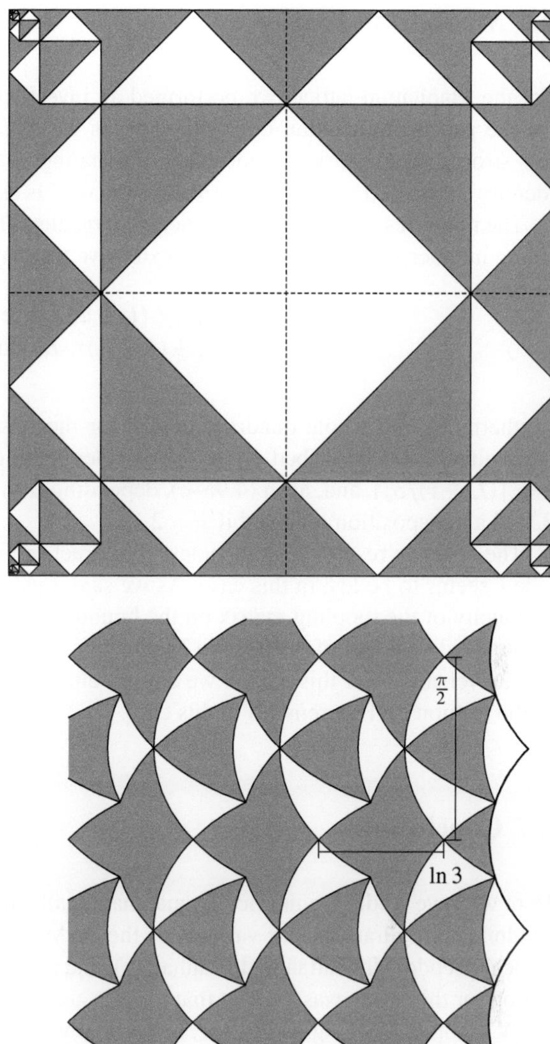

of the corner having a $b_{ij} = 1$ every two sites. This check must be performed only on the deterministic part, and again is done in a straightforward way, or, with conceptual economy, implied by the explicit numerical check on 6 consecutive sizes. This completes the proof of all the three theorems. □

4.6 Manhattan Lattice

For the Manhattan lattice, we performed an investigation similar to the one above
for the pseudo-Manhattan one, although, as we already said, the motivations are
less strong, as this lattice is not planar alternating. A small-size example of Creutz
identity, compared to the pseudo-Manhattan one, is shown in Fig. 4.8.

The numerics gave positive and negative results. The positive results concern the
filling number k_L, that, according to extensive tests (up to $L \simeq 100$) we conjecture
to be

$$k_L = \begin{cases} \frac{1}{4} L(L + 2) & L \text{ even} \\ \frac{1}{4} (L + 1)^2 & L \text{ odd} \end{cases} \tag{4.21}$$

Furthermore, the whole quadrant except for the $\ell \times (\ell - 1)$ corner seems to be
deterministically described by a set of rules analogous to Eq. (4.9a–e) (again,
$\ell = \lfloor (L + 1)/3 \rfloor$), and, as in (4.9a–e), depending from the congruence of L modulo
3, and a transposition involved if $s = 2$.

The negative result is that the telescopic exact self-similarity between side L and
side ℓ seems to be lost in this case. As we said, this property relies crucially on the
simplicity of the toppling matrix on the boundary of the quadrants, which seems to
be an accidental fact of the pseudo-Manhattan lattice, and has few chances to show
any universality. For this reason we did not attempt to state and prove any theorem
in the fashion of Theorem 4.6 in this case.

4.7 Conclusions

Here we have studied, numerically and analytically, the shape of the Creutz identity
sandpile configurations, for variants of the ASM, with directed edges on a square
lattice (pseudo-Manhattan and Manhattan), and square geometry. An original motivation for this study was the fact that heights are valued in $\{0, 1\}$ (while the original
BTW sandpile has heights in $\{0, 1, 2, 3\}$), and we conjectured that this could led to
simplifications. The results have been even simpler than expected, and qualitatively
different from the ones in the BTW model.

In the BTW model, the exact configuration seems to be unpredictable: although
some general "coarse-grained" triangoloid shapes seem to have a definite large-
volume limit, similar in the square geometry and in the one rotated by $\pi/4$, here and
there perturbations arise in the configuration, along lines and of a width of order 1 in
lattice spacing. We discuss the role of these structures in various aspects of the ASM
in the paper [13] and in Chap. 5.

The triangoloids have precise shapes depending from their position in the geometry, and are smaller and smaller towards the corners. Understanding analytically at
least the limit shape (i.e. neglecting all the sub-extensive perturbations of the regular-
pattern regions) is a task, at our knowledge, still not completed, although some first

important results have been obtained in [19]; further achievements in this direction have come with the work of Levine and Peres [20, 21], both in the similar context of understanding the relaxation of a large pile in a single site (see in particular the image at page 10 of Levine thesis, and the one at
http://www.math.cornell.edu/~levine/gallery/invertedsandpile1m10x.png,
and, for the directed model, the one at page 20 of
http://www.newton.kias.re.kr/~nspcs08/Presentation/Dhar.pdf).

In our Manhattan-like lattices on square geometry, however, we show how the situation is much simpler, and drastically different. Triangoloids are replaced by exact triangles, all of the same shape (namely, shaped as half-squares), and with straight sides. All the sides of the triangles are a fraction $\frac{1}{2}3^{-k}$ of the side of the lattice (in the limit), where the integer k is a "generation" index depending on how near to a corner we are, and indeed each quadrant of the configuration is self-similar under scaling of a factor $1/3$. The corresponding "infinite-volume" limit configuration is depicted in Fig. 4.9. A restatement of the self-similarity structure, in a language resembling the $z \to 1/z^2$ conformal transformation in Ostojic [19] and Levine and Peres [20], is the fact that, under the map $z \to \ln z$, a quadrant of the identity (centered at the corner) is mapped in a quasi–doubly-periodic structure.

Also the filling numbers (i.e. the minimal number of frame identities relaxing to the recurrent identity) have simple parabolic formulas, while in the original BTW model a parabola is not exact, but only a good fitting formula.

These features reach the extreme consequences in the pseudo-Manhattan lattice, where the exact configuration at some size is deterministically obtained, through a ratio-1/3 telescopic formula. These facts are not only shown numerically, but also proven directly in a combinatorial way.

In Chap. 5 and in the article [13], we describe how it is possible to achieve a complete comprehension of the shape of generic triangoloids that appear in many deterministic protocols of the ASM, see in particular Sect. 5.6.

References

1. D. Dhar, T. Sadhu, S. Chandra, Pattern formation in growing sandpiles. EPL **85**, 48002 (2009)
2. S. Caracciolo, G. Paoletti, A. Sportiello, Explicit characterization of the identity configuration in an abelian sandpile model. J. Phys. A Math. Theor. **41**, 495003 (2008). arXiv:0809.3416v2
3. M. Creutz, Abelian sandpiles. Comp. in Phys. **5**, 198–203 (1991)
4. M. Creutz, Abelian sandpiles. Nucl. Phys. B (Proc. Suppl.) **20**, 758–761 (1991)
5. M. Creutz, Playing with sandpiles. Phys. A Stat. Mech. Appl. **340**, 521–526 (2004). Complexity and Criticality: in memory of Per Bak (1947–2002)
6. D. Dhar, P. Ruelle, S. Sen, D.N. Verma, Algebraic aspects of abelian sandpile models. J. Phys. A Math. Gen. **28**, 805 (1995)
7. M. Creutz, P. Bak, Fractals and self-organized criticality, in Fractals in science, ed. by A. Bunde, S. Havli(Springer, Berlin 1994), pp. 26–47
8. R. Cori, D. Rossin, On the sandpile group of dual graphs. Eur. J. Comb. **21**, 447–459 (2000)
9. S.N. Majumdar, D. Dhar, Height correlations in the abelian sandpile model. J. Phys. A Math. Gen. **24**, L357 (1991)

10. S.N. Majumdar, D. Dhar, Equivalence between the abelian sandpile model and the $q \to 0$ limit of the potts model. Phys. A Stat. Mech. Appl. **185**, 129–145 (1992)
11. Y. Le-Borgne, D. Rossin, On the identity of the sandpile group. Discrete Math. **256**, 775–790 (2002). LaCIM 2000 Conference on Combinatorics, Computer Science and Applications.
12. M. Creutz, Xtoys: cellular automata on xwindows, Nucl. Phys. B Proc. Suppl. **47**, 846–849 (1996). code: http://thy.phy.bnl.gov/www/xtoys/xtoys.html
13. S. Caracciolo, G. Paoletti, A. Sportiello, Conservation laws for strings in the abelian sandpile model. EPL **90**, 60003 (2010). arXiv:1002.3974v1
14. S. Caracciolo, M.S. Causo, P. Grassberger, A. Pelissetto, Determination of the exponent γ for saws on the two-dimensional manhattan lattice. J. Phys. A Math. Gen. **32**, 2931 (1999)
15. J.T. Chalker, P.D. Coddington, Percolation, quantum tunnelling and the integer hall effect. J. Phys. C Solid State Phy. **21**, 2665 (1988)
16. T. Sadhu, D. Dhar, Pattern formation in growing sandpiles with multiple sources or sinks. J. Stat. Phy. **138**, 815–837 (2010). doi:10.1007/s10955-009-9901-3
17. D. Dhar, T. Sadhu, Pattern formation in fast-growing sandpiles. ArXiv e-prints, (2011), arXiv:1109.2908v1
18. T. Sadhu, D. Dhar, The effect of noise on patterns formed by growing sandpiles, J. Stat. Mech. Theory Exp. **2011**, P03001 (2011). arXiv:1012.4809
19. S. Ostojic, Patterns formed by addition of grains to only one site of an abelian sandpile. Phys. A Stat. Mech. Appl. **318**, 187–199 (2003)
20. L. Levine, Y. Peres, Scaling limits for internal aggregation models with multiple sources. J. d'Anal. Mathé. **111**, 151–219 (2010). doi:10.1007/s11854-010-0015-2
21. L. Levine, Limit theorems for internal aggregation models, Ph.D. thesis, University of California, at Berkeley, fall 2007 arXiv:0712.4358. http://math.berkeley.edu/~levine/levine-thesis.pdf

Chapter 5
Pattern Formation

Strings, Backgrounds and their Classification

It has been spent a huge amount of efforts in the study of Abelian Sandpile Model, as prototype of Self Organized Criticality. These studies have mainly dealt with the critical exponents of avalanches produced in sandpiles driven slowly in their critical steady state through a stochastic dynamics. Recently the attention on the sandpile model has grown also in other aspects, different from the study of critical exponents. Indeed in particular deterministic protocols has been possible to see beautiful and intriguing patterns with shapes and defects.

In recent works of Dhar and collaborators [1–4] is studied the emergence of patterns in some given periodic configurations after the addition of sand on a single site; they study also the case of multiple sites, the effect of the presence of sinks and of noise, and finally they determine the asymptotic shapes in some interesting cases. The studies presented take place first on the standard undirected square lattice, which is the space of definition of the classic BTW model, but a complete formula determining the detailed shapes of the patterns is given only for particular lattices called the F-lattice (which we investigate under the name of pseudo-Manhattan lattice in Chap. 4 when studying the recurrent identity) and the Manhattan lattice.

Here we investigate some *experimental* protocols on the standard square lattice based on the action of the operators Π_i introduced in Chap. 3, these protocols produce a full family of *patches* and *strings*, over a number of backgrounds. The relation between these objects are given and their classification is fully given through a recursive procedure. Afterwards we present a protocol to produce Sierpiński triangoloids whose fundamental elements are these patches and strings, these triangoloids are the key structure arising in the classification of strings and backgrounds.

Part of the results here presented are published in the article [5].

G. Paoletti, *Deterministic Abelian Sandpile Models and Patterns*,
Springer Theses, DOI: 10.1007/978-3-319-01204-9_5,
© Springer International Publishing Switzerland 2014

5.1 Introduction

While the main structural properties of the ASM can be discussed on arbitrary graphs [6], for the subject at hand here we shall need some extra ingredients (among which a notion of translation), that, for the sake of simplicity, suggest us to concentrate on the original realization on the square lattice [7], within a rectangular region $\Lambda \in \mathbb{Z}^2$.

So in this setting, we write $i \sim j$ if i and j are first neighbors. The configurations are vectors $z \equiv \{z_i\}_{i \in \Lambda} \in \mathbb{N}^\Lambda$ (z_i is the number of sand-grains at vertex i). Let $\bar{z} = 4$, the degree of vertices in the bulk, and say that a configuration z is *stable* if $z_i < \bar{z}$ for all $i \in \Lambda$. Otherwise, it is *unstable* on a non-empty set of sites, and undergoes a relaxation process whose elementary steps are called *topplings*: if i is unstable, we can decrease z_i by $\bar{z}$, and increase z_j by one, for all $j \sim i$. The sequence of topplings needed to produce a stable configuration is called an *avalanche*.

Avalanches always stop after a finite number of steps, which is to say that the diffusion is strictly *dissipative*. Indeed, the total amount of sand is preserved by topplings at sites far from the boundary of Λ, and strictly decreased by topplings at boundary sites. The stable configuration $\mathcal{R}(z)$ obtained from the relaxation of z, is univocally defined, as all valid stabilizing sequences of topplings only differ by permutations.

We call a stable configuration *recurrent* if it can be obtained through an avalanche involving all sites in Λ, and *transient* otherwise (Sect. 3.1.2). Recurrent configurations have the structure of an Abelian group (Sect. 2.2) under the operation $z \oplus w := \mathcal{R}(z + w)$. We have only a partial knowledge of the group identity for each Λ (see e.g. [8, 9]; recently a complete characterization has been achieved for a simplified directed lattice, the *F-lattice* [10]) nonetheless they are easily obtained on a computer and they provide a first example of the intriguing complex patches in which we are interested. The maximally-filled configuration $z_{\max}$, with $(z_{\max})_i = \bar{z} - 1 = 3$ for all i and density equal to 3, is recurrent. More generally, for large Λ, recurrent configurations must have average density $\rho(z) = |\Lambda|^{-1} \sum_i z_i \geq 2 + o(1)$ (this bound is tight). So structures with density $\rho > 2$, $\rho = 2$ and $\rho < 2$ are said respectively *recurrent*, *marginal* and *transient*.

We call *patch* a region filled with a periodic pattern. The *density* ρ of a patch is the average of z_i within a unit tile. Neighboring patches may have an interface, periodic in one dimension, along a vector which is principal for both patches. Let us suppose that in a deterministic protocol [1] we generate a region filled with polygonal patches, glued together with such a kind of interfaces. At a vertex where $\ell \geq 3$ patches meet, label cyclically with $\alpha = 1, \dots, \ell$ these patches, call ρ_α the corresponding densities, and θ_α the angles of the interfaces between the patch α and $\alpha + 1$ (subscripts $\alpha = \ell + 1 \equiv 1$). These quantities are proven to satisfy the relation in $\mathbb{Q} + i\mathbb{Q}$

$$\sum_{\alpha=1}^{\ell} (\rho_{\alpha+1} - \rho_\alpha) \exp(2i\theta_\alpha) = 0 \tag{5.1}$$

which has non-trivial solutions only for $\ell \geq 4$ [1].

Fig. 5.1 A string with momentum $(6, 1)$, in a background pattern with periodicities $V = ((2, 1), (0, 2))$. String and background unit cells are shown in *gray*. The density in the string tile is $\rho = (18 \cdot 3 + 8 \cdot 2 + 4 \cdot 1 + 7 \cdot 0)/(6^2 + 1^2) = 2$

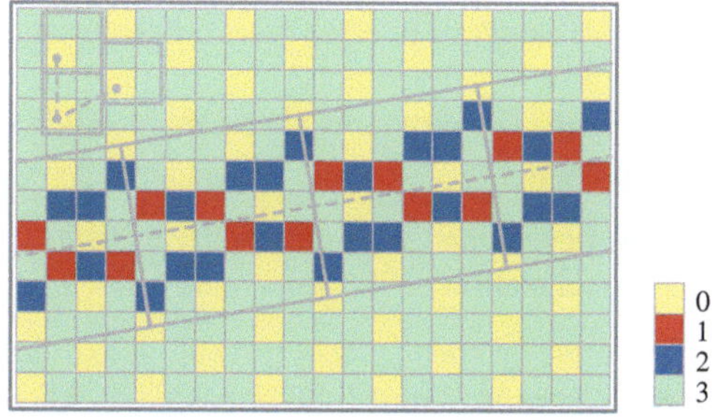

We call *string* a one-dimensional periodic defect line, with periodicity vector $k = (k_x, k_y) \in \mathbb{Z}^2$[1], that we call *momentum*, in a background patch, periodic in both directions, and has k as a periodicity vector. The background on the two sides may possibly have a periodicity offset (see Fig. 5.1).

5.2 Experimental Protocols for Strings and Patches

Consider a two-dimensional lattice Λ, with vertex-set isomorphic to $\mathbb{Z}^2$, and edge-set sharing the periodicity of $\mathbb{Z}^2$, so that each vertex has $\bar{z}$ neighbors in Λ. (The exclusion, made here, of the possibility of an internal basis, as e.g. in the honeycomb lattice, is done only in order to keep a light notation.) Take as toppling rules $\vec{\Delta}_i$ and threshold parameters $\bar{z}_i$ the ones induced by the lattice. Now consider a simply-connected domain $\Omega \in \mathbb{R}^2$. Then, for ϵ sufficiently small, consider the ASM model on $\mathcal{G}_\epsilon := \Omega \cap \epsilon\Lambda$. Various portions of the boundary may have *open* or *closed* boundary conditions. A site $\vec{n} \in \Omega \cap \epsilon\Lambda$ with b neighbors out of Ω has $\bar{z}_{\vec{n}} = \bar{z}$ and $b_{\vec{n}} = b$ in case of open conditions, and $\bar{z}_{\vec{n}} = \bar{z} - b$ and $b_{\vec{n}} = 0$ in case of closed conditions. The system dissipates mass at the open portion of the boundary of Ω, or in more and more far regions of Ω, if the latter is not compact.

Then, call an *experimental protocol* $\mathfrak{P}$ a configuration $z(\epsilon)$ obtained through the deterministic evolution of a given (ϵ-dependent) initial condition $w(\epsilon)$ on $\mathcal{G}_\epsilon$.

For pairs $(\vec{x}, \epsilon) \in \Omega \times \mathbb{R}^+$, define some function $\vec{x}'_\epsilon(\vec{x}) \in \mathcal{G}_\epsilon$ such that $|\vec{x}'_\epsilon - \vec{x}| = \mathcal{O}(\epsilon)$, and consider the *local coordinate system*, for $i \in \epsilon\Lambda$ and 'near' to $\vec{x}$: $i = \vec{x}'_\epsilon + \epsilon\vec{n}$, with $\vec{n} \in \mathbb{Z}^2$ and $|\vec{n}|\epsilon \ll 1$.

We say that the point $\vec{x}$ has a *weak (thermodynamic) limit* under protocol $\mathfrak{P}$, if a function $x'_\epsilon(x)$ as above exists such that, for ϵ arbitrarily small, a subset of $[0, \epsilon]$ with measure $\mathcal{O}(\epsilon)$ exists, in which the description of $z(\epsilon)$ through a local coordinate system near x is independent from ϵ, and determined by some explicit formula in $\vec{n}$.

We say that the point $\vec{x}$ has a *strong (thermodynamic) limit* under protocol $\mathfrak{P}$, if the same holds as above, except for the fact that the 'good' behavior should concern

---

[1] Here and in the following, bold letters $k, v, \dots$ are vectors in $\mathbb{Z}^2$ if not otherwise stated.

the whole interval. In other words, besides some value L, for *all* sizes the behavior of $z(\epsilon)$ near x is determined (instead of just a finite fraction).

A number of protocols have been considered in the literature, mostly for BTW, among them:

1. the determination of the recurrent identity [8, 9, 11].
2. the relaxation of $\lfloor \epsilon^{-2} \rfloor \delta_{i0} + u$, where u is a periodic configuration [1, 12, 13].

All these protocols, for various choices of lattice Λ and domain Ω, seem to show a 'weak' thermodynamic limit almost everywhere (and even a strong limit, but only in some regions and with trivial patterns). Instead, for comparison, the same protocols in a peculiar variant of the ASM on a directed square lattice (called F-lattice or Pseudo-Manhattan lattice) show a strong limit everywhere [1, 10].

The possibility of having at least a weak limit comes from the fact that the problem has the form of a discrete integer-valued Laplacian, i.e., for a given unstable configuration[2] w determined by the protocol, one has to find the stable configuration $z = \mathcal{R}(w)$ and the integer-valued vector T such that

$$\Delta T + z = w \, . \tag{5.2}$$

Remark that also the problem of determining the recurrent identity is in this class, as $Id_{\rm r} = \mathcal{R}(kb)$, as stated in (2.36), where b has a definite thermodynamic limit, and $k \sim \epsilon^{-2}$.

If it was not for the requirement that T is integer-valued (and its counterpart, the possibility for an offset z in the resulting configuration), the equation above would describe a simple Laplacian problem on a lattice, formally solved by methods of Green functions, for which even the perturbative series of lattice corrections is well understood. These new ingredients, being an aspect of the non-linearity of the problem, are responsible for the emerging interesting behaviors.

5.2.1 Master Protocol

Before introducing the details of the master protocol, we recall some notions on the ASM, introduced in Chap. 3. We have the sand-addition operators $\tilde{a}_i$'s and the toppling operators t_i's, as defined in Sect. 2.1.1, they are clearly invertible. So it is possible to introduce the sand-removing operators $\tilde{a}_i^{-1}$'s and the antitoppling operators t_i^{-1}'s, as done in Sect. 3.1.1 Eq. (3.5). As a consequence, besides the standard relaxation operator $\mathcal{R}$ which is defined by means of topplings, we have an *anti-relaxation* operator $\mathcal{R}^{\dagger}$, which maps $\mathbb{Z}^n$ to S_+, and S_- to S, by performing a sequence of antitopplings.

As a corollary, within the space of stable configurations, the toppling/antitoppling operators $\Pi_i = a_i^{\dagger} a_i$ act as projectors, translating the configurations inside their

---

[2] We use here the notation of Chap. 2

equivalence class. In each class there is a single stable configuration which is invariant under the application of any Π_i, result that we proven in Theorem 5.1.

We now define the *master* protocol (for definiteness in the BTW case and starting from the maximally filled configuration, but extensions are immediate). Take Λ to be the square lattice, and Ω to be a polygon with L sides of slope $\lambda_1, \ldots, \lambda_L$ and open boundary conditions. Consider a string of points $(\vec{x}_1, \ldots, \vec{x}_t)$ in Ω, and, for any ϵ, the corresponding nearest points in $\Omega \cap \epsilon \Lambda$.

Then, almost surely (w.r.t. Lebesgue measure in Ω^t), the configuration obtained acting with $\Pi_{x_t} \cdots \Pi_{x_1}$ on the maximally-filled configuration has a strong limit everywhere except that at the corners of the domain, and is constituted by a web of strings, in a maximally-filled background, meeting at scattering vertices (with momentum conservation), and obtained through a simple geometric construction, iterative in $t' = 1, \ldots, t$. Strings may end up on the boundary, without momentum conservation, but only in a neighborhood (of size related to t) of the corners. Top-left image in Figs. 5.2 and 5.5 are examples of such configurations.

This protocol explores the transient configurations on the given equivalence class $[z]$ of the recurrent representative. Indeed taking into account the action of a_i followed by $a_i^\dagger$ on a given configuration z an easy calculation shows that the two configurations are toppling equivalent. Call T the toppling vector associated to the action of a_i and U the one for $a_i^\dagger$ then we have

$$\Pi_i z = z - \sum_j \vec{\Delta}_j (T_j - U_j) \tag{5.3}$$

that is, the two configuration are equivalent under toppling and belong to the same class.

Let now describe in detail the action of the operators Π_i's on a particular example: consider a rectangular region Λ, and the maximally filled configuration $z_{\max}$. We split the action of Π_i in two parts, first the addition of the grain followed by the relaxation, then the removal followed by the anti-relaxation. So add one grain of sand at some vertex j and then relax the configuration, $\mathcal{R}(z_{\max} + e_i)$; the resulting configuration an inner rectangle, of strings $(1, 0)$ and $(0, 1)$, equidistant from the border of Λ and having j on its perimeter, the corners of this rectangle are connected to the corners of Λ with strings $(1, 1)$ and $(-1, 1)$. This picture has exactly one defect at j, manifested as a single extra grain, w.r.t. the underlying periodic structure, see Fig. 5.2. The configuration $\mathcal{R}(z_{\max} + e_i)$ is obviously recurrent. Now remove this extra grain in j and anti-relax the configuration, if necessary, the configuration is now transient. Repeat the procedure at some new vertex j', say in the region below the inner rectangle. In the resulting configuration new strings $(2, 1)$ and $(-2, 1)$ appear. Iterating this procedure, with Λ large enough, strings with higher and higher momenta are generated. Furthermore, given that the unit tiles of strings with different momenta are classified, this protocol is completely predictable, for arbitrary Λ, through a purely geometric construction.

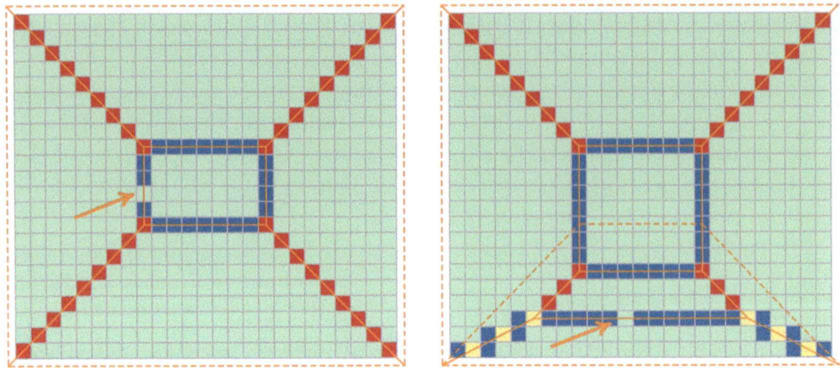

Fig. 5.2 On the *left*, the configuration obtained after relaxation from z_{max} plus an extra grain of sand exactly at the vertex where a defect appears. On the *right*, the result after removing the defect and the addition of one more grain

Briefly, in a given recurrent background, with translation vectors $V = (\boldsymbol{v}_1, \boldsymbol{v}_2)$, one and only one string of momentum $\boldsymbol{k} = m_1\boldsymbol{v}_1 + m_2\boldsymbol{v}_2 = \boldsymbol{m}V$ can be produced, if $\gcd(m_1, m_2) = 1$, and no strings of momentum $\boldsymbol{k}$ exists for $\boldsymbol{k}$ not of the form above. V and V' are equivalent descriptions of the background periodicity iff $V' = MV$, with $M \in \mathrm{SL}(2, \mathbb{Z})$. Accordingly, $\boldsymbol{m}' = \boldsymbol{m}M^{-1}$. Furthermore strings can meet in tri-vertices only if they satisfy the relation $\boldsymbol{k} = \boldsymbol{p} + \boldsymbol{q}$ with $\det \begin{pmatrix} p_x & p_y \\ q_x & q_y \end{pmatrix} = \boldsymbol{v}_1 \wedge \boldsymbol{v}_2$.

It is possible to generalize the master protocol following three directions, each one corresponding to changing of a particular feature of the starting configuration:

1. we can change the *slope* of the sides of the polygons;
2. we can change the *boundary conditions* of the boundary to be open or periodic;
3. we can change the *background* pattern of the configuration.

(1) When dealing with the master protocol on a general polygon Ω, the strings are created by the interaction of the avalanche started by Π_i with each side. Indeed when the operator Π_i acts on a site i in a recurrent patch lying on the border of Λ with slope $\boldsymbol{\ell}$, then a string of momentum $\boldsymbol{\ell}$ is detached from that side. The string is placed so that it intersects the point i.

This property allows to generate any string, with different slopes and in (almost) any point of Λ, simply changing the slopes of the sides and the application points. If a recurrent patch has two consecutive sides with slopes $\boldsymbol{\ell}_1$ and $\boldsymbol{\ell}_2$ on the open borders of λ, acting with Π_i generates both the strings of different slope, so they must intersect. In case they have $\boldsymbol{\ell}_1 \wedge \boldsymbol{\ell}_2 = 1$ they satisfy the scattering relations and meet in a scattering vertex generating the string $\boldsymbol{\ell}_3 = \boldsymbol{\ell}_1 + \boldsymbol{\ell}_2$. The creation of the vertex in terms of the single strings is explained in Sect. 5.5 and is shown in Figs. 5.4 and 5.20.

(2) An open boundary is essential in order to create the strings, when acting with the Π_i's. If we have periodic boundary conditions on two opposite sides, then they

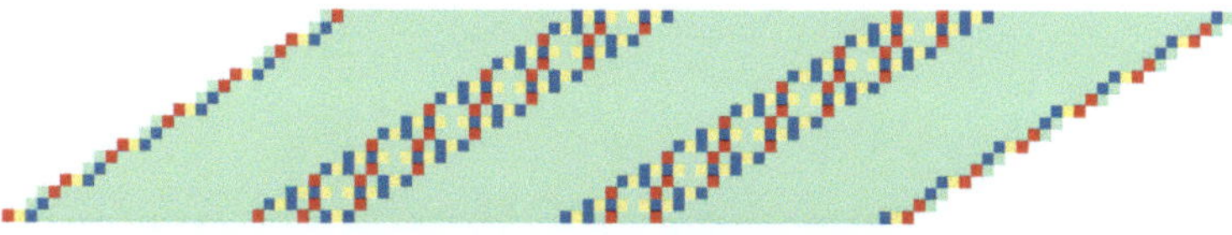

Fig. 5.3 Construction of the string $k = (5, 3)$ through the master protocol with slopes k and periodic boundary up and down

do not generate any string, in particular a string can cross them going from one of these sides to the other. This feature allows to generate one shot any string we want. It suffices to take the parallelogram with two opposite sides of slope k and two horizontal sides. Then we assign open boundary condition to the sides of slope k and periodic boundary conditions on the horizontal sides. The action of a single Π_i generates two strings of slope k, one crossing the point i and the second symmetric w.r.t. the axis parallel to k. These strings are "infinite" along the direction of the periodicity. An example of the result of such a protocol can be seen in Fig. 5.3 where the use of periodic boundary conditions allows to create, one shot, the desired string.

(3) We have presented the protocol starting from the maximally filled configuration, the first reason pushing us to use it is that the avalanches produced when acting with Π_i touch all the sites of the system, when we first start the protocol. However this configuration is not the only one, among recurrent periodic configuration, that displays this property. Other *background* configurations can be chosen as long as they are periodic, recurrent and allow avalanches touching all the sites V, the patches defined in Sect. 5.4 have these properties and will be studied in this framework, and there it will be elucidated the role of $z_{\max}$ as primitive background in the classification of these structures. Repeating then the same protocol, but acting with Π_i's only on sites where their action is not trivial[3], it is possible to produce a web of string, just as starting from $z_{\max}$, in any allowed background. We must check carefully the slopes of the polygon, indeed they have to correspond to some strings, thus being an integer linear combination of the vectors generating the background with coprime components. When strings meet in vertices, the value of the cross product of the incident strings has not to be 1 anymore but $v_1 \wedge v_2$ being v_1 and v_2 the generators of the translation for the background.

A final remark: we note that, using as starting configuration, not the maximally filled configuration or a configuration filled with a unique recurrent periodic patch, but a recurrent configuration with patches composed of the recurrent backgrounds defined in Sect. 5.4, then the action of Π_i on sites i with height $\bar{z}_i - 1$, still generates the web of strings as in the other cases. The difference arises when strings overlap a patch that was not maximally filled. In these zones the resulting string is perturbed, usually being the string in the maximally filled background decreased as it was decreased w.r.t. $z_{\max}$ the patch; and it comes back to normality when reaching the

---

[3] The sites we are interested in are the ones on the border of the framing polygons described in Sect. 5.4

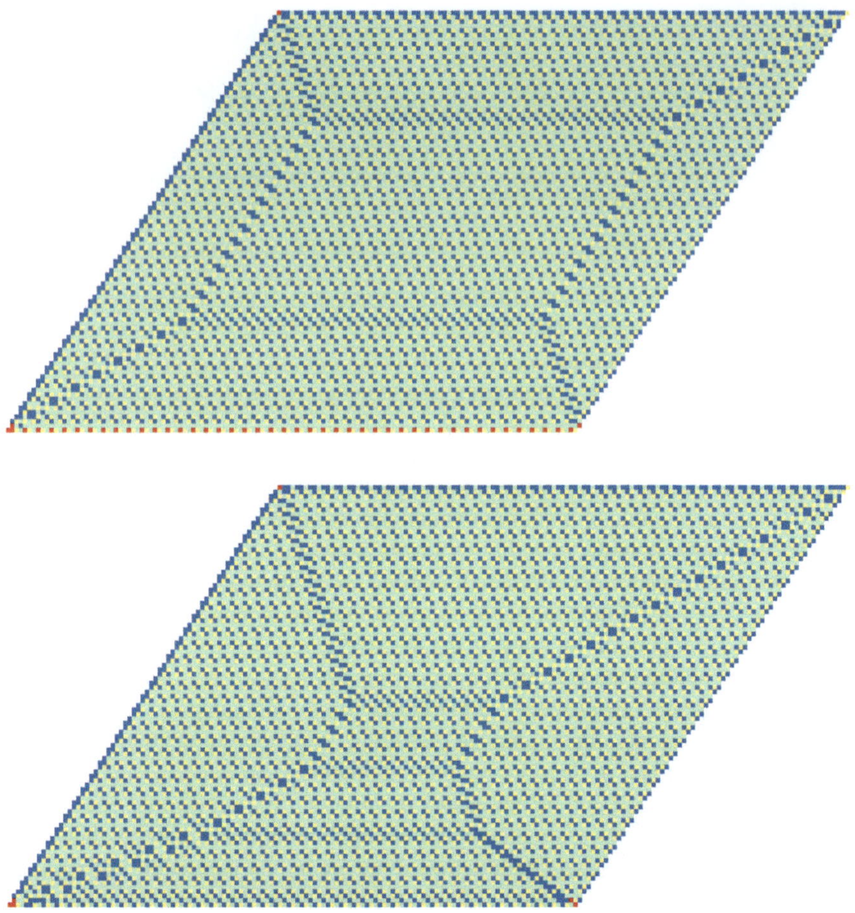

Fig. 5.4 Determination of scattering vertices through the master protocol with open boundaries in a rhomboidal geometry for a background $\Phi(3, 1)$, here we use the function (5.10) to identify a background from its associated string on the maximally filled background. The slopes are $p = (2, 3)$ and $q = (3, 0)$, which corresponds to the smallest horizontal vector linear combination of v_1 and v_2 of the background. The protocol generates 4 scattering vertices in the *top figure*, acting with Π_i on the perimeter of the *inner rhombus*. Successive action of the operator add a higher momentum string q such that the new scattering vertex $k' \to k + q$ adds to $k \to p + q$

maximally filled space, we call this effect the *Cheshire cat effect*. A scheme of what happen, and a concrete example, is given in Fig. 5.5.

5.2.2 Wild Orchids

In Sect. 3.4 we described an experimental protocol which is a Markov Process defined on the space of stable configurations S. Starting from the maximally filled

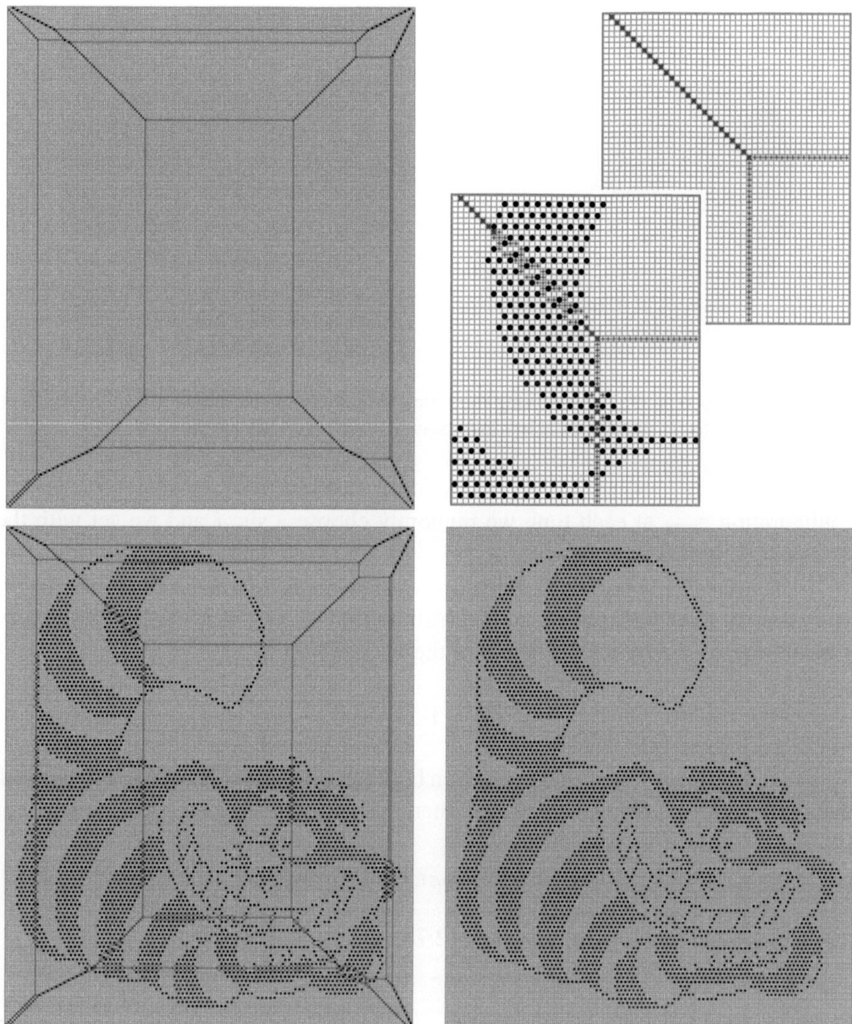

Fig. 5.5 Illustration of the *Cheshire cat effect*. *Top left* a realization of the master protocol, in a rectangular geometry $(L_x, L_y) = (200, 240)$ on the maximally filled background. The sequence of Π_i's positions, randomly chosen, is $\big((64, 85), (58, 212), (118, 96), (141, 145), (145, 223), (6, 102), (192, 95), (16, 201), (188, 89), (57, 30)\big)$. *Bottom right* an arbitrary recurrent drawing, such that the height is 3 in all the positions above (the image is a low-resolution reproduction of the artwork for the 1951 Disney animated film "Alice in Wonderland"). *Bottom left* after applying the same sequence of Π_i's to this configuration, it results in essentially the same geometric structure of network of strings, with suitable local modifications when the dark parts of the drawing are crossed. This modifications do not translate the positions of the strings, as if the cat had the property of being invisible to the master protocol. A magnified detail showing clearly this property is visible in the *top-right* images

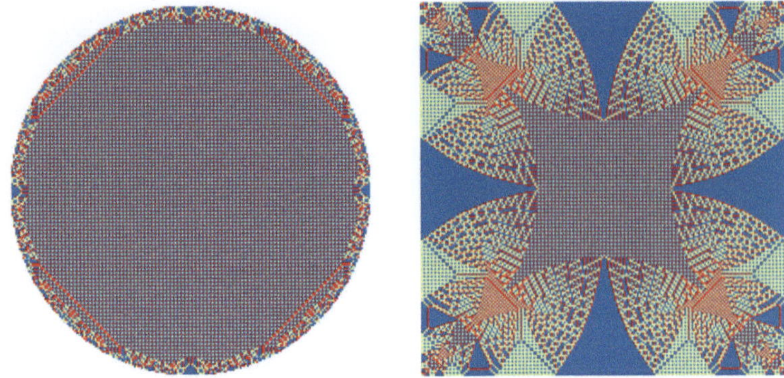

Fig. 5.6 The two Wild Orchids resulting from the convergence of the Markov Chain dynamics on a circle of diameter $D = 200$ on the *left* and a square of side $L = 200$ on the *right*

configuration $z_{\max}$ at each time we randomly choose a site i and we act with the idempotent combination $\Pi_i = a_i^\dagger a_i$.

In Theorem 5.1 is proved that this process converges to a fixed configuration, τ, that we have called *Wild Orchid* and is illustrated in Fig. 5.6 and in Fig. 3.1. Obviously τ is left unchanged by the the action of the Π_i's, which is

$$\Pi_i \tau = \tau \quad \forall i \in V \tag{5.4}$$

Theorem 5.1 makes use of the connection between this particular Markov Chain and the multitoppling rules, see Sect. 2.6, where we consider toppling cluster's set $\mathcal{L}$ of (2.77)

$$\mathcal{L} = \big\{ \{i, j\}_{ij \in E} \big\} \cup \big\{ \{i\}_{i \in V} \big\},$$

that induces the toppling rule given in (2.78)

$$\text{if } \begin{cases} ij \in E \\ z_i \geq \Delta_{ii} + \Delta_{ij} \\ z_j \geq \Delta_{jj} + \Delta_{ij} \end{cases} \implies z_k \to z_k - \Delta_{ik} - \Delta_{jk} \quad \forall k.$$

It is easy to see that the action of Π_i is trivial unless the site i is part of an unstable cluster E in the sense of (2.78). So acting with Π_i correspond to a local change of the toppling rule, as described in Sect. 3.3.

5.3 First Results

In this section we report about the results obtained in the investigations carried out with the help of the methods introduced in the previous Section.

In a given recurrent background, with translation vectors $V = (v_1, v_2)$, one and only one string of momentum $k = m_1 v_1 + m_2 v_2 = mV$ can be produced, if $\gcd(m_1, m_2) = 1$, and no strings of momentum k exists for k not of the form above. V and V' are equivalent descriptions of the background periodicity iff $V' = MV$, with $M \in \mathrm{SL}(2, \mathbb{Z})$. Accordingly, $m' = mM^{-1}$. And indeed, sets of $m \in \mathbb{Z}^2$ with given gcd's are the only proper subsets invariant under the action of $\mathrm{SL}(2, \mathbb{Z})$. The gcd constraint arises in the classification of the elementary strings, because, when $d = \gcd(k_x, k_y) > 1$, the corresponding periodic ribbon is just constituted of d parallel strings with momentum k/d.

The unit tile of each string, as well as of each patch, is symmetric under 180-degree rotations. In particular, momenta k and $-k$ describe the *same* string. The tile of a string of momentum k fits within a square having k as one of the sides, so that each string is a row of identically filled squares. This is a non-empty statement: the tile could have required rectangular boxes of larger aspect ratio, and even an aspect ratio depending on momentum and background.

A string of momentum k has an *energy E*, defined as the difference of sand-grains, in the framing unit box of side k, w.r.t. $z_{\max}$. We have the relation $E = |k|^2$, or, in other words, the unit tile has exactly marginal density, $\rho = 2$, irrespectively of the density of the surrounding background (as seen, e.g., in Fig. 5.1).

Two strings, respectively of momentum p and q, can collapse in a single one of momentum k (see Fig. 5.7). In this process momentum is *conserved*: $p + q = k$. More precisely, the strings join together in such a way that the square boxes surrounding the unit cells meet at an extended *scattering vertex*, a triangle of sides of lengths equal to $|k|$, $|p|$ and $|q|$, rotated by $\pi/2$ w.r.t. the corresponding momenta: given this geometrical construction, momentum conservation rephrases as the oriented perimeter of the triangle being a closed polygonal chain.

Local momentum conservation and the $k \leftrightarrow -k$ symmetry are reminiscent of equilibrium of tensions, in a planar network of tight material strings, from which the name.

On networks, this local conservation is extended to a global constraint. Choose an orthogonal frame (x, t), and orient momenta in the direction of increasing t. Then, sections at fixed t are all crossed by the same total momentum. Rigid extended domain walls between periodic patterns, satisfying similar local and global conservations, appear in certain tiling models [14–16], which remarkably show a Yang-Baxter integrable structure, where the corresponding strings are usefully interpreted as worldlines of particles in the (x, t)-frame. Note, however, that, at variance with these models, in the ASM we have an infinite tower of excitations, for a given background, and infinitely many different backgrounds too.

In the maximally-filled background, because of the (D_4 dihedral) symmetry, we can restrict without loss of generality to study strings of momentum k with both components positive. For each such k with $\gcd(k_x, k_y) = 1$, simple modularity reasonings show that there exists a *unique* ordered pair of momenta p and q, with non-negative components, such that $p + q = k$ and the matrix $\begin{pmatrix} p_x & p_y \\ q_x & q_y \end{pmatrix}$ is in $\mathrm{SL}(2, \mathbb{Z})$, see Appendix A for more details on its structure. We write in this case

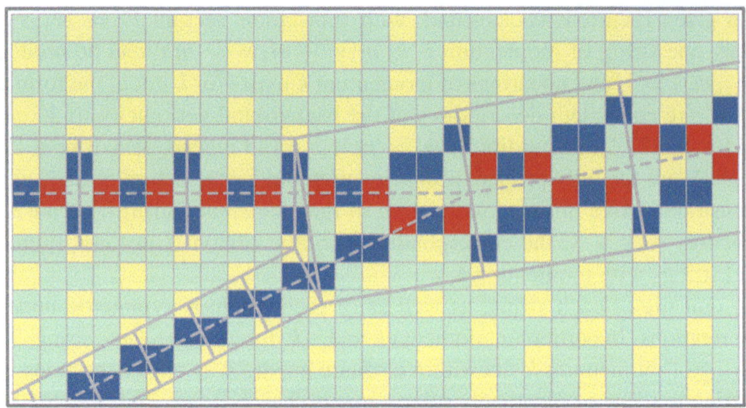

Fig. 5.7 A scattering involving strings with momenta $(4, 0)$, $(2, 1)$ and $(6, 1)$, on the background pattern of Fig. 5.1 (also symbol code is as in Fig. 5.1)

$k \leftarrow (p, q)$. For example, $(10, 3) \leftarrow \big((7, 2), (3, 1)\big)$. The endpoint of p, starting from the top-left corner of the k framing box, is the (unique) lattice point which is nearest to the top-side of the box. This alternative definition generalizes to non-trivial backgrounds, and the $m_1 v_1 + m_2 v_2$ sublattice.

Let us go back to the problem of ℓ interfaces which meet at a given corner, but allow now, besides interfaces between patches, incident strings. Following the analysis of [1], and therefore using the graph-vector $T = \{T_i\}$, where T_i is the number of topplings at i in the relaxation of the starting configuration, we study its characteristics in a region that, in the starting configuration, was uniformly filled with a patch. However, now we allow for toppling distributions which are piecewise both quadratic and linear (the linear term was neglected in [1], as subleading in the coarsening).

For any relevant direction α, allow for a patch interface, or a string, or both. Call $\tilde{E}^{(\alpha)}$ the difference for *unit length* (not for period), in the total number of grains of sand w.r.t. $z_{\max}$, due to presence of a string, i.e. $\tilde{E}^{(\alpha)} = E^{(\alpha)}/|k^{(\alpha)}|$, or the contribution from a non-zero impact parameter in the interface. It can be shown, by reasonings as in [1], that the difference between the extrapolated toppling profile for two contiguous patches, at a polar coordinate (r, θ), must be of the form

$$T_{r,\theta}^{(\alpha+1)} - T_{r,\theta}^{(\alpha)} = \frac{r^2}{2}\,(\rho_{\alpha+1} - \rho_\alpha)\,\sin^2(\theta - \theta_\alpha)$$
$$+ r\,\tilde{E}^{(\alpha)}\sin(\theta - \theta_\alpha) + \mathcal{O}(1)\,. \tag{5.5}$$

Then, by summing over α and matching separately the quadratic and linear terms, we conclude that, for each θ,

Fig. 5.8 From *left* to *right*, the recurrent, marginal and transient patches constructed from the propagator $k = (3, 2) \leftarrow \big((2, 1), (1, 1)\big)$ in the maximally filled background, having densities $\rho = 2$ and $2 \pm 1/12$ (symbol code is as in Fig. 5.1)

$$\begin{cases} \sum_{\alpha=1}^{\ell} (\rho_{\alpha+1} - \rho_\alpha) \sin^2(\theta - \theta_\alpha) = 0 \\ \sum_{\alpha=1}^{\ell} \tilde{E}^{(\alpha)} \sin(\theta - \theta_\alpha) = 0 \end{cases} \tag{5.6}$$

so that, besides the anticipated equation (5.1) for patches alone, that was deduced in [1], we obtain

$$\sum_{\alpha=1}^{\ell} \tilde{E}^{(\alpha)} \exp(i\theta_\alpha) = 0 \tag{5.7}$$

which describes the string and interface-offset contributions.

In (5.1), the first non-trivial value for ℓ is 4 [1]. In our generalization, 4 is the minimal value for the number of patches *plus* the number of strings, and thus includes new possibilities: a *scattering* event, with three incident strings in a single background, as in Fig. 5.7, and the case of two strings and two patches, producing diagrams reminiscent of total *reflection* and *refraction* in optics, so that the specialization of (5.7) can be read as a *Snell's law* for ASM strings. For the case of three strings on a common background $\mathcal{B}$, we get

$$\sum_{\alpha=1}^{\ell} \frac{E^{(\alpha)}}{|k^{(\alpha)}|^2} k^{(\alpha)} = 0 \tag{5.8}$$

which shows that momentum conservation implies a dispersion relation of the form $E^{(\alpha)} = c_{\mathcal{B}}|k^{(\alpha)}|^2$, and viceversa.

The classification of the strings preludes to a classification of the patches. To any string of momentum k, univocally decomposed as $k \leftarrow (p, q)$, we can associate three patches, respectively recurrent, marginal and transient, through a geometrical construction, involving p and q, sketched in an example in Fig. 5.8, a full construction procedure is given in the next Section.

Reflection and refraction events also appear. Let us have a triple k, p and q such that $k \leftarrow (p, q)$, then consider a single string of momentum $k' = mp + q$ (for m a large integer), and in the scattering of $mp + q$ into q and m parallel p strings.

A consequence of the recursive construction of the string textures, the string of momentum k' is forced to look as a strip-shaped patch of m-period width, of the marginal tile associated to p, crossed by a 'soft' string, that reflects twice per period k', up to ultimately leaving the marginal patch, through a refraction, and propagates in the recurrent background.

We will show in the next section how the interplay between strings and patches, both at the level of classification and of evolution in deterministic protocols, is the key-ingredient to clarify allometry in pattern formation for the ASM, and to design new protocols in which short-scale defects are totally absent. The resulting structure is a fractal, a Sierpiński triangoloid whose structure is elucidated in Sect. 5.6, where the theoretical formula (5.1) has infinitely many distinct realizations.

5.4 Patches and Strings on the $\mathbb{Z}^2$ Lattice

Consider $\mathbb{Z}^2$ and the ASM defined on its vertices, a pictorial representation of the model is obtained considering the dual lattice, which is obviously itself, and filling each plaquette with a color corresponding to the height of its relative point in the direct lattice, an example of such a representation is given in Fig. 5.1 where it is shown the legend color↔height.

Let define $\mathcal{P}$ as the space of all possible *backgrounds*, or *patches* and Σ as the space of all possible *strings*; then $\Sigma = \bigcup_{B \in \mathcal{P}} \Sigma_B$, where Σ_B is the space of all possible strings on a given background $B \in \mathcal{P}$.

We call $B \in \mathcal{P}$ a *background* if it is a height configuration which corresponds in the dual lattice to a tiling of $\mathbb{Z}^2$ with wallpaper group 2222 and regular hexagonal tiles, with borders on the lattice which identify a contour of height 3. A wallpaper group is a type of topologically discrete group of isometries of the euclidean plane that contains two linearly independent translations[4]. These translations identify a lattice, in such a way that the lattice points, corresponding to all the possible translations, are in the form $\{mv_1 + nv_2 \mid m, n \in \mathbb{Z}\}$ with v_1 and v_2 fixed vectors. It is now clear that the translation symmetry can be specified by two vectors, nevertheless the choice of these vectors is not unique, in fact any pair[5] $(av_1 + bv_2, cv_1 + dv_2)$ with $ad - bc = \pm 1$ creates the same lattice. We choose v_1 and v_2 in such a way that the vectors v_1, v_2 and $-v_1 - v_2$ determine an acute triangle, and we call these the *canonical generators* (in exceptional cases there could be an ambiguity and v_1, v_2 and $-v_1 - v_2$ originate a triangle with $\pi/2$ angle). A background is fully determined by the following objects:

- the *framing polygon $F(B)$*, which is the border of the tile;
- the *filling* of the framing polygon $H(B)$, the height function inside the tile;

---

[4] For more details see appendix D.

[5] Each pair also define a parallelogram, whose area is constant, equal to the magnitude of the cross product.

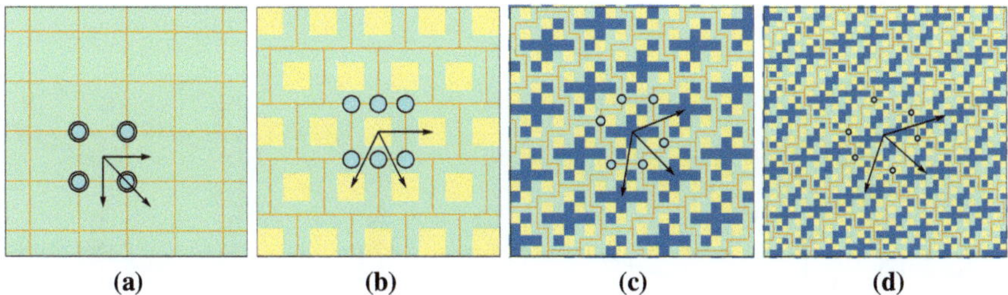

Fig. 5.9 Examples of backgrounds with different properties highlighted. Framing polygons are drawn with *yellow lines*, triple points are represented as *blue dots* and vectors v_1, v_2 and $-v_3$ are showed in every case. In (**a**) is shown the maximally filled background, it is a limit case where the triple points are at the intersection of four framing polygons and they are only 4 per tile, while in the typical case, they are 6 per tile and at the intersection of 3 of them, as it is in (**b**), (**c**) and (**d**)

- the *translation vectors* v_1, v_2[6];
- the *triple points*, 6 for every tile, placed on the intersection point of three adjacent tiles.

These parameters for a given background are not independent, e.g. the triple points can be obtained directly from $F(B)$ together with the translation vectors, furthermore we will say briefly *tile* understanding it as the framing polygon together with the filling. Furthermore we call the size of a background the volume of its framing polygon, which is $|F(B)|$.

We call $P \in \Sigma$ a *string* if it is a one-dimensional periodic defect line on a given background, that is described by the frieze group 22∞. It is fully determined by the following objects:

- the *habitat background* $B(P)$, where the string is settled;
- the *periodicity vector* $k(P)$, that identifies also a *framing box* of sides $k(P)$ and $ik(P)$;
- the *framing polygon* $F(P)$, which is the shape of the border of the tile producing the defect;
- the *filling* $H(P)$, the height function of the string's tile;
- the *blue cells*, points with height 2 placed at the connection between the framing polygons of the background and of the string so that the height 3 contour from opposite sides of the string become disconnected[7].

The difference on tiles' size[8] between habitat background and strings allows to distinguish two different types of strings. *Type I* strings are those with $|F(P)| < |F(B(P))|$ and *type II* string those with $|F(P)| > |F(B(P))|$. Type I strings have blue cells placed at the connection points between a tile of the string and

---

[6] In the following we will use v_3 in addition to v_1 and v_2 with the constraint $v_1 + v_2 + v_3 = 0$.

[7] The name blue cells comes from the color for sites of height 2 used in our representation.

[8] We call *tile size* the area of the corresponding framing polygon and we denote it by $|F(\cdot)|$.

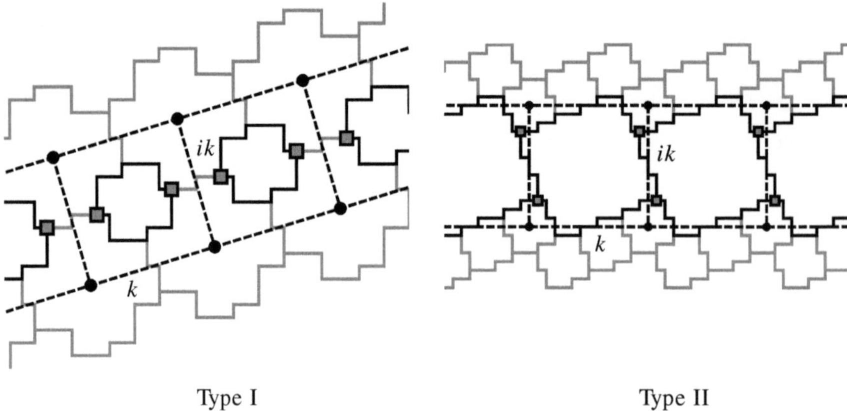

Type I   Type II

Fig. 5.10 *Type I* and *Type II* strings, in *gray* are the framing polygons of the background, in *black* the framing polygons of the strings, *black dots* are the points from which to draw the framing box, in *dashed lines*, and *gray squares* are the *blue* cells

two of the background, taken from the two opposite sides of the string and respectively at position $i\boldsymbol{k}$. Type II strings have blue cells at the connections between a tile of the background and two of the string. Furthermore type I strings correspond to the generators of the background, hence they are only three, their momentum being $\boldsymbol{k} = \boldsymbol{v}_\alpha$. The *framing box* for a string has vertices in the center of the four corner background tiles of the string, they sit on the square of sides $\boldsymbol{k}(P) \times i\boldsymbol{k}(P)$.

The spaces $\mathcal{P}$ and Σ are graduated spaces (on $\mathbb{N}$) with $|F(B)|$ and $|F(P)|$. In this framework, the maximally filled configuration (see Fig. 5.9) is the fundamental background, and it has $|F(B)| = 1$, moreover in all lattices the maximally filled configuration is a valid background and has the smallest possible $|F(B)|$. Then the construction of the strings in this background, and the duality background-strings that will be discussed later (5.16), allow to fill the list of all possible backgrounds and strings.

Let us call $L(v_1, v_2) = \mathrm{span}\{v_1, v_2\} \equiv \{(m_1 v_1 + m_2 v_2) \mid m_1, m_2 \in \mathbb{Z}\}$. If a string P has habitat B, and B has translation vectors $\boldsymbol{v}_1$ and $\boldsymbol{v}_2$, then $\boldsymbol{k}(P) \in L(\boldsymbol{v}_1, \boldsymbol{v}_2)$ and its component m_1 and m_2 are coprime, so we define a subset of $L(\boldsymbol{v}_1, \boldsymbol{v}_2)$ where the strings actually live, $L_\Sigma(\boldsymbol{v}_1, \boldsymbol{v}_2) \equiv \{(m_1 v_1 + m_2 v_2) \mid m_1, m_2 \in \mathbb{Z}, \ |m_1, m_2 \ \mathrm{coprime}\}$. We define the function:

$$\mathbb{A} : \Sigma \to L_\Sigma(\boldsymbol{v}_1, \boldsymbol{v}_2) \times L_\Sigma(\boldsymbol{v}_1, \boldsymbol{v}_2)$$
$$P \to (\boldsymbol{p}, \boldsymbol{q}) \tag{5.9}$$

that maps a string P into an ordered pair $(\boldsymbol{p}, \boldsymbol{q}) \in L_\Sigma(\boldsymbol{v}_1, \boldsymbol{v}_2) \times L_\Sigma(\boldsymbol{v}_1, \boldsymbol{v}_2)$ such that $\boldsymbol{p} + \boldsymbol{q} = \boldsymbol{k}(P)$ and $\boldsymbol{p} \wedge \boldsymbol{q} = \boldsymbol{v}_1 \wedge \boldsymbol{v}_2$ where $\boldsymbol{v}_1$ and $\boldsymbol{v}_2$ are the translation vectors of $B(P)$. The existence and unicity of this pair is provided by properties of $SL(2, \mathbb{Z})$ (this is proved in lemma 1 at page 131 in Appendix Sect. A.1). The existence of this function allows to associate an element $A = \begin{pmatrix} m_x\boldsymbol{p} & m_y\boldsymbol{p} \\ m_x\boldsymbol{q} & m_y\boldsymbol{q} \end{pmatrix} \in SL(2,\mathbb{Z})$ to every string, in such a way that the periodicity vector of the corresponding string is given by $\boldsymbol{k}(A) = (1,1)A\begin{pmatrix} \boldsymbol{v}_1 \\ \boldsymbol{v}_2 \end{pmatrix}$, where $\boldsymbol{v}_1, \boldsymbol{v}_2$ are generators of the background.

We define the function

$$\Phi : \Sigma^{(\nu)} \to \mathcal{P}^{(\nu)}$$
$$P \to B \qquad\qquad (5.10)$$

that maps a string P into the associated background B such that:

– $F(B) = F(P)$;
– $H(B) = H(P)$.

The translation vectors have to be chosen differently if P is a type I or a type II strings. In fact for type I strings we have:

– $\boldsymbol{v}_1(B) = i\boldsymbol{k} = i\boldsymbol{w}_\alpha$, where $\boldsymbol{w}_\alpha$ where $\alpha = 1, 2, 3$ are the translation vectors of the habitat background;
– $\boldsymbol{v}_2(B) = i(\boldsymbol{w}_\beta - i\boldsymbol{k})$, where $\beta \neq \alpha$ and the sign in front of $i\boldsymbol{k}$ is negative if $\boldsymbol{k} \wedge \boldsymbol{w}_\beta$ is positive, positive otherwise[9].

While for type II strings we have:

– $\boldsymbol{v}_1(B) = i\boldsymbol{k}$;
– $\boldsymbol{v}_2(B) = i(\boldsymbol{p} + i\boldsymbol{k})$, where $\boldsymbol{p}$ is given by $\mathbb{A}(P)$.

These properties completely determine the background. Applying Φ to a type I string the $\boldsymbol{v}_1$ and $\boldsymbol{v}_2$ given through this procedure are not the canonical generators, however the canonical generators belong to $L(\boldsymbol{v}_1, \boldsymbol{v}_2)$, and are the suitable pair that preserves the wedge product and generates an acute triangle. Clearly different strings can lead to the same background (see Fig. 5.11), only three of these strings are of type II and the others of type I.

We define the functions

$$\hat{\Psi}_{1,2,3} : \mathcal{P}^{(\nu)} \to \Sigma^{(\nu')} \quad (\nu' < \nu)$$
$$B \to P_{1,2,3} \qquad\qquad (5.11)$$

that return the three fundamental type I strings with habitat background $B(P_{1,2,3}) = B$. The strings P_α have periodicity vectors $\boldsymbol{v}_\alpha$ and framing polygons $F(P_\alpha)$ obtained as the empty spaces resulting when two strips of $F(B)$ with slope $\boldsymbol{v}_\alpha$ are translated one with respect of the other of $i\boldsymbol{v}_\alpha$. Furthermore given $P_\alpha \in \Sigma^{(\nu')}$ with $\nu' < \nu$ and being $F(P_\alpha)$ and the triple points already known, then $H(P_\alpha)$ is also known, since the hierarchical construction of Σ. In this process the limit case in

---

[9] We want $|\boldsymbol{v}_1(B) \wedge \boldsymbol{v}_2(B)| = |F(P)|$ and we know $|F(P)| = |\boldsymbol{k}|^2 - |F(B(P))|$, with $|F(B(P))| = \boldsymbol{w}_1 \wedge \boldsymbol{w}_2$. So we have $|\boldsymbol{v}_1(B) \wedge \boldsymbol{v}_2(B)| = i\boldsymbol{k} \wedge i(\boldsymbol{w}_\beta \pm i\boldsymbol{k}) = \boldsymbol{k} \wedge \boldsymbol{w}_\beta \pm \boldsymbol{k} \wedge \boldsymbol{k} = \boldsymbol{k} \wedge \boldsymbol{w}_\beta \pm |\boldsymbol{k}|^2$, the last two summands have to subtract each other, but $|\boldsymbol{k}|^2$ is always positive, so its sign has to change as the sign of $\boldsymbol{k} \wedge \boldsymbol{w}_\beta$ changes.

Fig. 5.11 Background corresponding to a framing polygon of large size $|F(B)| = 160$, that is a $\Phi(P)$ for the three strings of Fig. 5.12

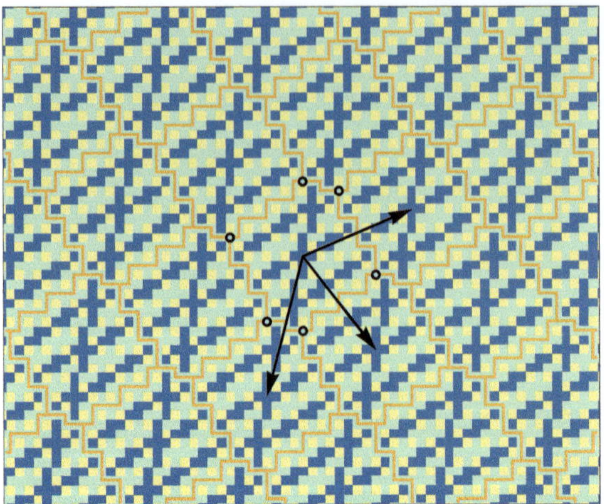

which $|F(P_\alpha)| = 0$ can occur, this is the case when $F(P)$ is square shaped, indeed it exists a periodicity vector v_α parallel to a side of the square (perpendicular to the others) and the procedure to obtain $F(P_\alpha)$, of the relative string, results in a zero size framing polygon, however there is no problem in placing the blue cells.

These strings are limit cases, and they do not correspond to any background through Φ, when needed we will read this situation as a lack of background.

We introduce the functions

$$\hat{\Phi}_{1,2,3} : \mathcal{P}^{(v)} \to \Sigma^{(v)}$$
$$B \to P_{1,2,3} \tag{5.12}$$

that return the three type II strings associated to a framing polygon of a given background. They are given by the following properties:

– $F(P_{1,2,3}) = F(B)$;
– $H(P_{1,2,3}) = H(B)$;
– $k_{1,2,3} = iv_{1,2,3}(B)$;
– $B(P_{1,2,3}) = \Phi(\Psi_{1,2,3}(B))$;

In the case of zero size $\Psi_\alpha(B)$ then the corresponding P_α has no habitat background and do not exist as proper type II string, although it appears in the Sierpiński construction, where the lack of background means it is placed at the border of the ASM configuration. We note that $\hat{\Phi}_\alpha$ and Φ satisfy the following relation

$$\Phi\hat{\Phi}_\alpha = \mathbb{I}_{\mathcal{P}} \tag{5.13}$$

Moreover we observe that, given $B' = \Phi\hat{\Psi}_\alpha(B)$ for some $B \in \mathcal{P}$ then it exists a string P with B' as habitat background such that $\Phi(P) = B$. Similarly, given

Fig. 5.12 Three different type II strings corresponding to the same framing polygon

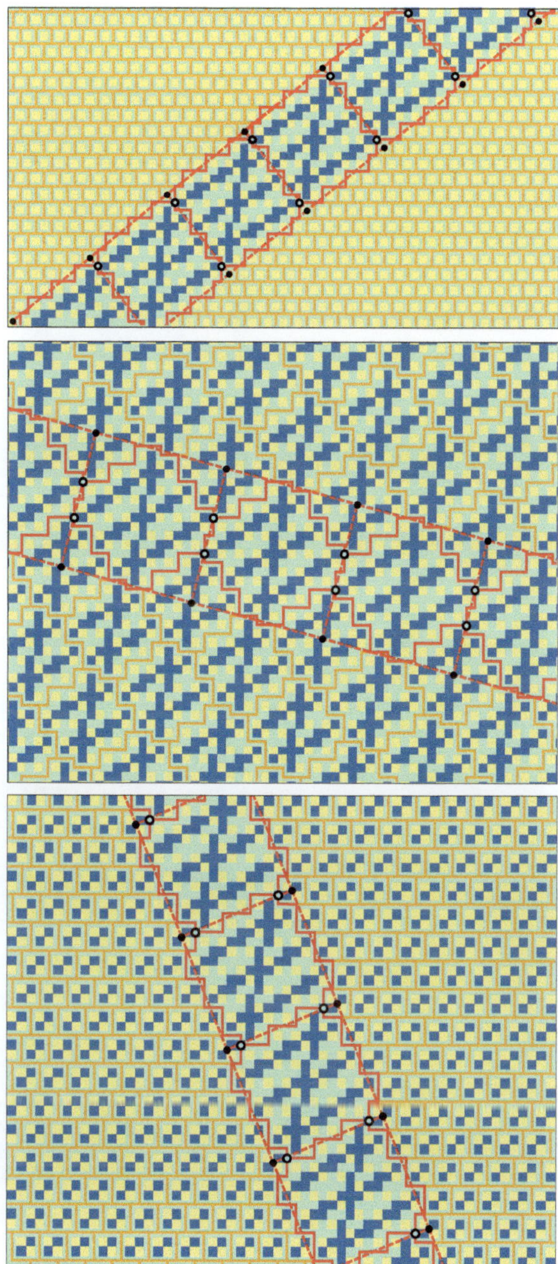

Fig. 5.13 The three fundamental strings on the background shown in Fig. 5.11

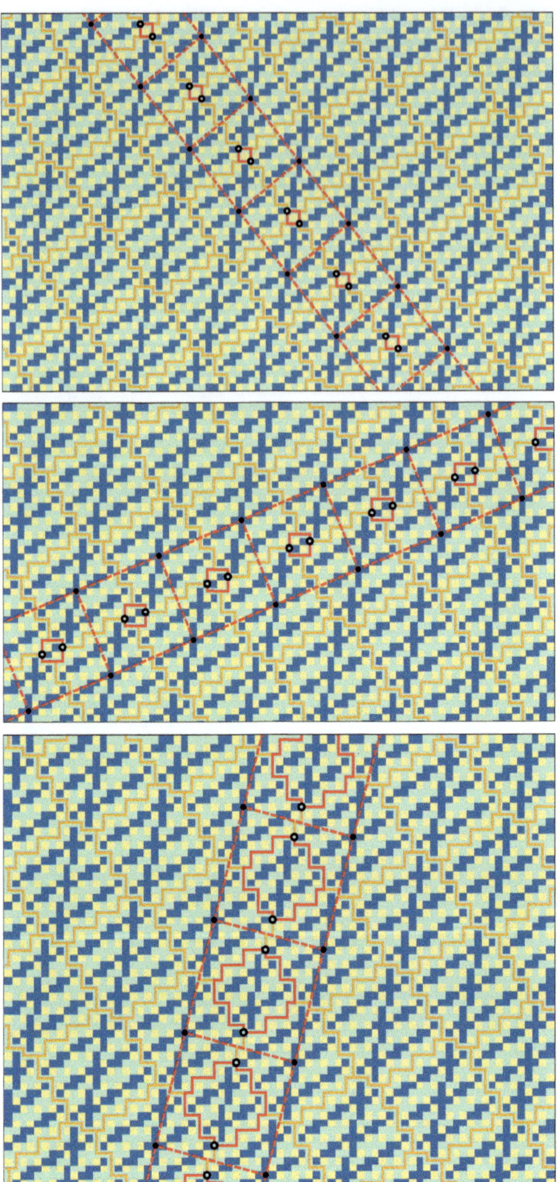

$P' = \hat{\Psi}_\alpha \Phi(P)$ for some $P \in \Sigma$ then P is a string on the background associated to P' through Φ.

We define the functions

$$\begin{aligned} \Psi_{1,2,3} : \Sigma^{(\nu)} &\to \mathcal{P}^{(\nu')} \quad (\nu > \nu') \\ P &\to B_{1,2,3} \end{aligned} \tag{5.14}$$

which return the three backgrounds where P is a type II string. This function is defined by

$$\Psi_\alpha = \Phi \hat{\Psi}_{1,2,3} \Phi \tag{5.15}$$

In other words, this means that, given a string P, the fundamental type I strings of the associated background through (5.10) generate the backgrounds where P lives as a type II string.

We have now all the ingredients to prove the duality connection which exists between strings and backgrounds in the following theorem.

Theorem 1 *Let us take two couples $(P, B(P))$ and $(P', B(P'))$ in $\Sigma \times \mathcal{P}$ such that the size of their tiles is greater than zero and with $B(P') = \Phi(P)$ and P' such that $F(P') = F(B(P))$. Then it exists a duality relation between them, which is made explicit by the following function:*

$$\begin{aligned} \mathcal{J} : \quad \Sigma \times \mathcal{P} &\to \Sigma \times \mathcal{P} \\ (P_0, B_0) &\to (P_1, B_1) \end{aligned} \tag{5.16}$$

with $B_0 = B(P_0)$ and $B_1 = B(P_1)$. The action of $\mathcal{J}$ is given as follow:

- $B_1 = \Phi(P_0)$
- $k(P_1) = ik(P_0)$
- $F(P_1) = F(B_0)$
- $H(P_1) = H(B_0)$

Proof Given the definition of $\mathcal{J}$, (P_0, B_0) has a type I string iff (P_1, B_1) has a type II string (and viceversa), this is clear cause to the interchange of framing polygons at each use of $\mathcal{J}$ and the definition of type I or type II strings in relation to the comparison of sizes of $F(P)$ and $F(B(P))$.

To complete the statement $\mathcal{J}$ being a duality relation we need to check that $\mathcal{J}$ is an involution, that is $\mathcal{J}^2 = \mathbb{I}$. The action of $\mathcal{J}^2$ can be summarized as follow:

$$(P_0, B_0) \xrightarrow{\mathcal{J}} (P_1, B_1) \xrightarrow{\mathcal{J}} (P_2, B_2) \tag{5.17}$$

If $(P_2, B_2) = (P_0, B_0)$ then $\mathcal{J}^2 = \mathbb{I}$.

At a level of framing polygons and fillings, the identity is trivial. In fact given the actions of Φ and $\mathcal{J}$, the tiles are exchanged twice so they do not actually change from (B_0, P_0) to (B_2, P_2).

It suffices now to check the equality for the periodicity vectors of the strings and the translation vectors of the backgrounds in order to complete the construction. The periodicity vector of the string is multiplied by i at each step, resulting in a final minus sign[10], so $P_0 = P_2$.

It remains to analyze the translation vectors of the background. We distinguish two cases: P_0 a type I or a type II string.

Let us first consider the case of P_0 type II string, then the relation (5.9) gives the decomposition $k_0 = p + q$ and $B_1 = \Phi(P_0)$, so in particular $v_1(B_1) = ik_0$ and $v_2(B_1) = i(p+ik_0)$, now $B_2 = \Phi(P_1)$, P_1 is a type I string and $v_1(B_1) \wedge v_2(B_1) > 0$[11], therefore $v_1(B_2) = i(ik_0) = -k_0$ and $v_2(B_2) = i(ip - k_0 - i(ik_0)) = -p$. This two vectors, $-k_0$ and $-p$ are in $L(v_1(B_0), v_2(B_0))$ by definition and their wedge product is $-k \wedge -p = (p+q) \wedge p = q \wedge p = v_1(B_0) \wedge v_2(B_0)$ so they generate the same background of $v_1(B_0)$ and $v_2(B_0)$ that is $B_2 = B_0$.

If instead P_0 is a type I string we can assume without loss of generality $k_0 = v_1(B_0)$ and we have $k_0 \wedge v_2(B_0) = v_1(B_0) \wedge v_2(B_0) > 0$. So $B_1 = \Phi(P_0)$, with $v_1(B_1) = ik_0$ and $v_2(B_1) = i(v_2(B_0) - ik_0)$. P_1 is a type II string, but given the translation vectors of the background it has a type I description, in fact $k_1 = ik_0 = v_1(B_1)$, and $v_1(B_1) \wedge v_2(B_1) < 0$, now $B_2 = \Phi(P_1)$ and in particular $v_1(B_2) = -k_0 = -v_1(B_0)$ and $v_2(B_2) = i(i(v_2(B_0) - ik_0) + i\,ik_0) = -v_2(B_0)$, so $B_2 = B_0$, and $\mathcal{J}$ is an involution. $\square$

The whole demonstration process is sketched in Table 5.1.

Examples of couples $(P, B) \leftrightarrow \mathcal{J}(P, B)$ are given in the images of Fig. 5.12 in correspondence with the ones of Fig. 5.13. $\mathcal{J}$ is not defined for zero size string, although they can be associated with particular structures sitting on the size of the configuration, and in this sense having no background, this discussion will be resumed in Sect. 5.6.

We finally remark the unique properties of the maximally filled background $B_{\max}$ picked up together with its string $(1, 1)_{B_{\max}}$ (indeed $(1, 0)_{B_{\max}}$ and $(0, 1)_{B_{\max}}$ have zero size and are not object of the theorem), the crucial point is the equality between tiles' sizes of the string and the background, 1. Cause to this equality the couple $((1, 1)_{B_{\max}}, z_{\max})$ is self-dual which is

$$(1, 1)_{B_{\max}}, B_{\max}) = \mathcal{J}((1, 1)_{B_{\max}}, B_{\max}). \tag{5.18}$$

This property will be a key in the understanding of self-similarity of particular Sierpiński structures in Sect. 5.6.

---

[10] k and $-k$ indentify the same string.

[11] $v_1(B_1) \wedge v_2(B_1) = ik_0 \wedge i(p + ik_0) = (p+q) \wedge p + k_0 \wedge k_0 = v_1(B_0) \wedge v_2(B_0) + |k_0|^2 > 0$

Table 5.1 Scheme of the action of $\mathcal{J}$ on strings of type I and type II

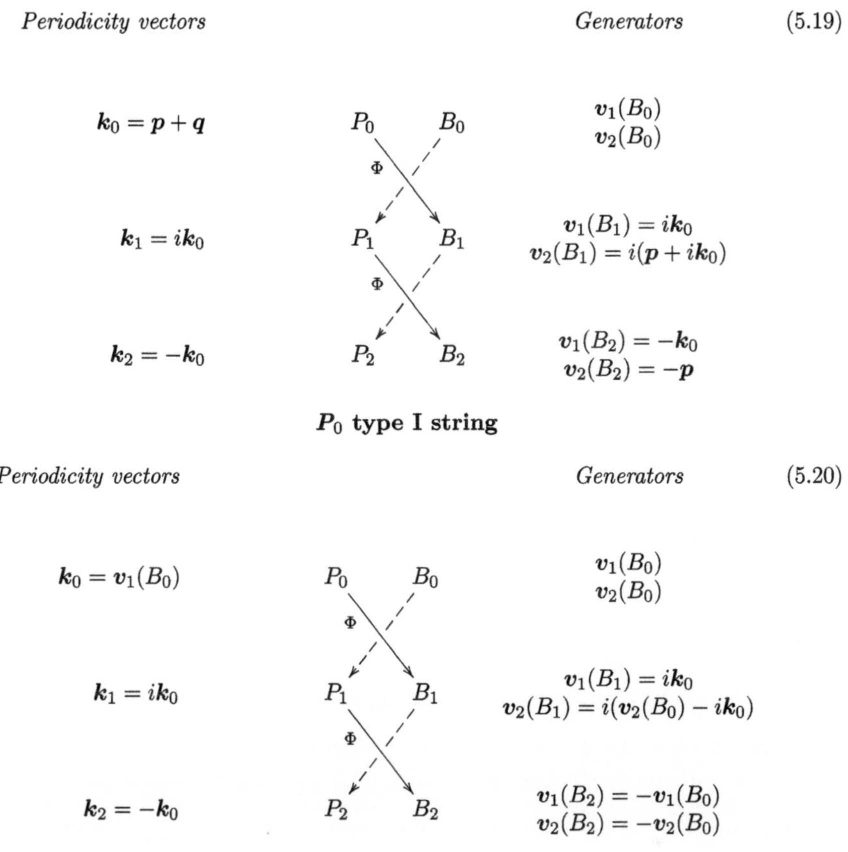

5.5 Strings Construction and Vertices

Given a background B, its space of strings Σ_B is composed of the three fundamental type I strings given by $\hat{\Psi}_\alpha(B)$ and a full list of type II strings with periodicity vectors in $L_\Sigma(v_1, v_2)$. As $L_\Sigma(v_1, v_2)$ is generated by two among v_1, v_2 and v_3, being this a redundant set of gerators, then Σ_B is generated through a recursive procedure by the three $\hat{\Psi}_\alpha(B)$. We focus our discussion on a fixed background B and we denote each string only by its periodicity vector.

The three v_1, v_2 and v_3 naturally divide the plane in three sectors, delimited by $(v_\alpha, v_{\alpha+1})$, we take $v_4 = v_1$. Cause to the 2-fold rotational symmetry saying that v and $-v$ identify the same string, and being $v_3 = -v_1 - v_2$, each sector can be further divided so that we have 6 sub-sectors, (v_α, v_β). Thus in each sector (v_α, v_β) we have

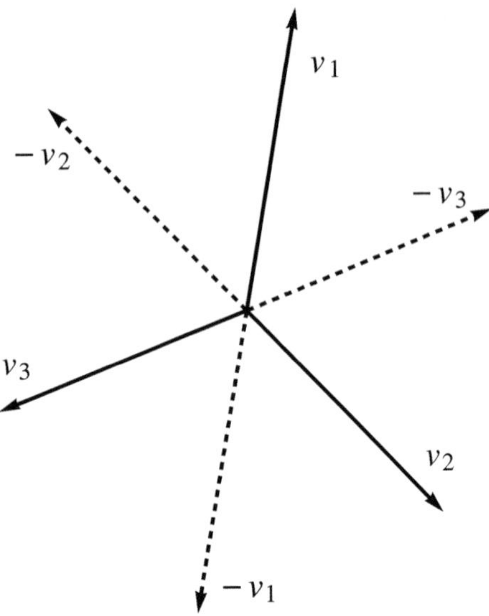

Fig. 5.14 Division of the plane by means of generators v_1, v_2 and v_3 considering the 2-fold rotational symmetry, which identifies v and $-v$

a subset of the type two strings $L_\Sigma(v_\alpha, v_\beta)$, with slope between the two vectors, that can be expressed in term of v_α and v_β). Indeed each string k is expressed through (5.9) as sum of p and q with $|p|, |q| < |k|$; the subsequent application of $\mathbb{A}$ to the vectors obtained step by step creates a succession of decreasing strings that finishes when fundamental strings are reached.

Procedure 2 *($k \leftarrow (p, q)$) We describe how higher momentum strings k are built in terms of their components $(p, q) = \mathbb{A}(k)$, on a given habitat background.*

Reasoning in terms of framing boxes, the framing box of the string, $k \times ik$, can be tessellated by means of sub-framing boxes $p \times ip$ and $q \times iq$; let us call A, B, C, D the vertices of $k \times ik$ as in Fig. 5.15, then $p \times ip$ sits on D and B while $q \times iq$ on A and C. This construction yields to some region with definition problems, i.e. regions covered by different framing boxes or regions uncovered. Being $q \neq p$ it is $|q| > |p|$ (or vice versa), then the two q framing boxes overlap in a central region of size roughly $|q - p|$. The recursive construction of q itself as $q \leftarrow (p, q - p)$ forces the overlap region to be the same.

Recursively this procedure allows to express the framing box of k in terms of framing boxes of the fundamental strings, a problematic overlap region remains, as well as uncovered regions of size $O(\sqrt{|F(B)|})$, the white regions in Fig. 5.15.

The formulation of this same construction in terms of framing polygons and height functions, that we shortly call *tile*, solves these problems. Let start by identifying the four background tiles at corners of $k \times ik$, and labeling them A, B, C, D as we did for the framing boxes; now we place the framing polygons for p and q starting from B, D and A, C respectively, just as if they were usual strings. Now look at the just formed sides of k (AB and DC), the construction forces the two different strings

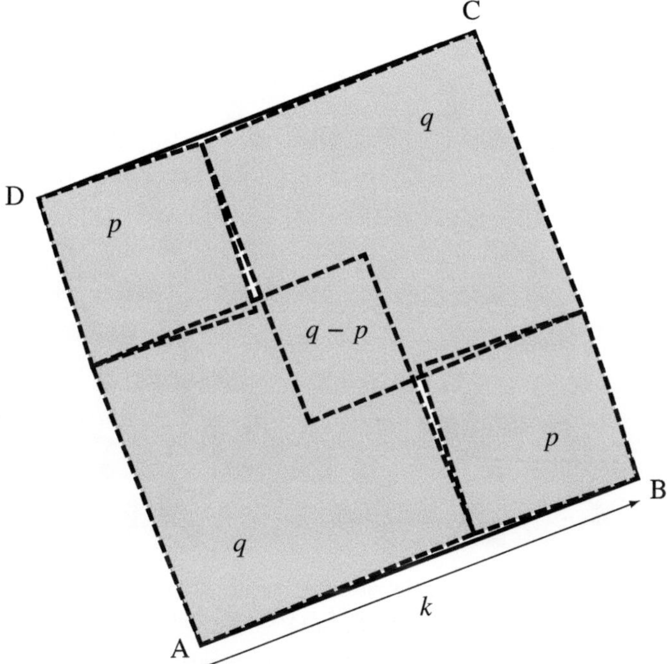

Fig. 5.15 Construction of $k \leftarrow (p, q)$ by means of framing boxes, framing boxes $p \times i\,p$ and $q \times i\,q$ are the *dashed lines*

departing on $A(D)$ and $B(C)$ to end in the same background tile, without any gap. At this point the tile for k is completely tessellated, we need only to give a prescription to place the blue cells. Where p and q face the background tiles, the standard blue cells are placed; in addition to these, the symmetric of the corners blue cells with respect to the center of sub-framing polygon are positioned. The overlap region appearing in this construction corresponds now exactly to the framing polygon of $q - p$, and all the blue cells already placed are kept.

The tile of k is the connected inner region once deleted the corners blue cells; this region is composed of two p tiles, two q tiles and two background tiles, connected by the blue cells. An example of this construction is shown in Fig. 5.16 and in Fig. 5.18.

Strings, on a given background, can interact in scattering vertices. In a vertex three strings meet together following strict rules and satisfying the relation (5.7). A vertex $k \rightarrow p + q$ must be such that the momentum, or in other word the periodicity, is conserved; so it is $k = p + q$, furthermore they have to be such that $p \wedge q = v_1 \wedge v_2$ such a triple of vectors is given for any allowed string k by $\mathbb{A}(k)$. The minimal triple of periodicity vectors satisfying this request is v_1, v_2 and $-v_3$, these correspond to the three fundamental strings $\hat{\Psi}_\alpha(B)$ of the background. The vertex corresponding to the minimal triple is the *fundamental scattering vertex*, in Fig. 5.17 it is shown for the $\Phi(2, -5)$ background where it is $(5, 2) \rightarrow (1, 6) + (4, -4)$.

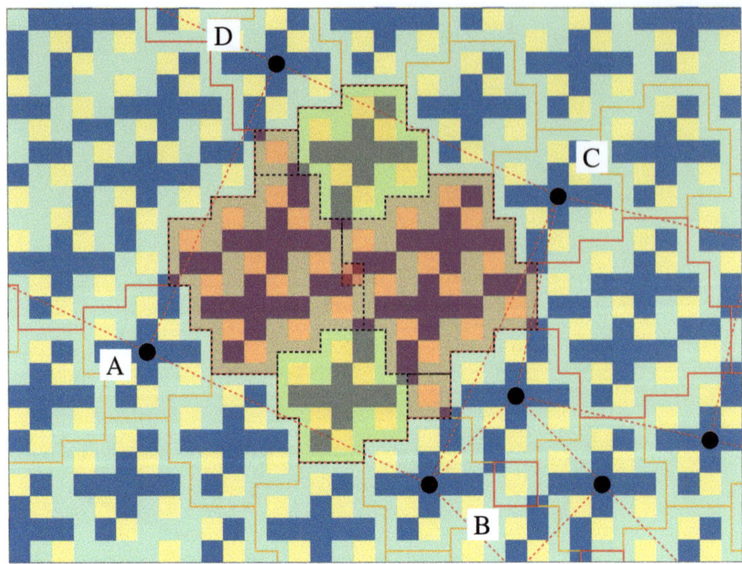

Fig. 5.16 Construction of $k \leftarrow (p, q)$ by means of framing polygons, it is shown also the vertex $k \rightarrow p+q$. Here $p = (4-4), q = (9, -2)$ and $k = (13, -6)$, the habitat background is $\Phi(2, -5)$

Fig. 5.17 The fundamental scattering in the $\Phi(2, -5)$ background, $(5, 2) \rightarrow (1, 6)+ (4, -4)$

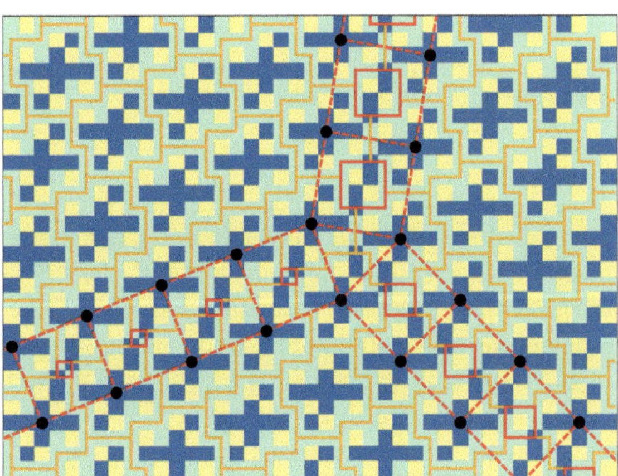

Higher momentum vertices are completely described by means of tiles, in particular, the procedure used to explain the construction $k \leftarrow (p, q)$ is the key to predict pointwise the shape of the vertex.

Procedure 3 $(p + q \rightarrow k)$ *We describe how to construct the scattering vertex in a process* $p + q \rightarrow k$ *where the strings satisfy* $(p, q) = \mathbb{A}(k)$.

First, we note that, reasoning by framing boxes, naturally emerges a vertex triangle whose sides are the three ik, ip and iq one for each scattering string, the area of this

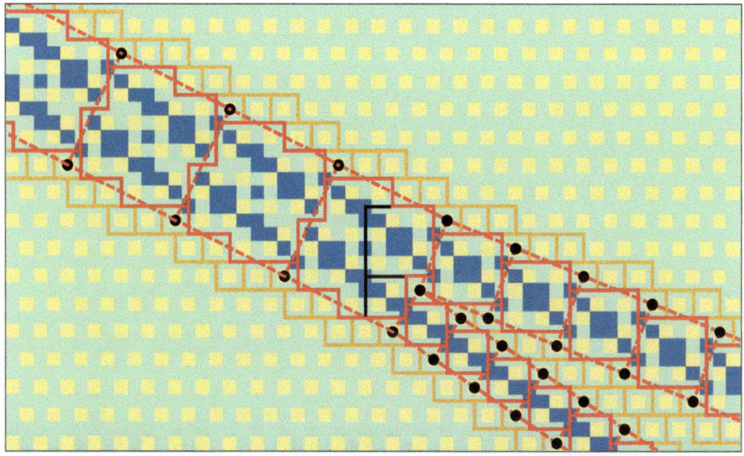

Fig. 5.18 Scattering vertex $(8, -4) \rightarrow (5, -2) + (3, -2)$ in the background $\Phi(2, -1)$

triangle $\frac{1}{2} v_1 \wedge v_2$, given the condition on the wedge product on p and q. The strings are defined up to the framing box connected to the vertex triangle, thus leaving part of the area of the triangle undefined. Let us suppose without loss of generality to have a vertex $k \rightarrow p + q$ with $|k| > |q| > |p|$ and $k \leftarrow (p, q)$ through (5.9), then k has the two p and q tiles closing the framing polygon at points B, C as in Fig. 5.16; their position perfectly fits with a string p departing from B and a string q departing from C. This construction places the outgoing strings in the proper order of increasing slope, given $p \wedge q > 0$. The blue cells placement is straight forward, indeed it suffices to place all the blue cells for every string connected to the vertex; so the two background tiles corresponding to the triangle vertices adjacent to ik have a single blue cell each one, in the usual position for k, p and q respectively, the tile connecting ip and iq has two blue cells corresponding to the strings placed, and this complete the construction.

This construction shows how the fundamental scattering vertex is characterized by having a minimal perimeter vertex triangle, which is equal to the sum of the length of the strings participating in the vertex among which the choice v_1, v_2 and v_3 is the minimum; the structure of the vertex is obtained placing three background tiles at iv_1, iv_2 and iv_3 one with respect of the other. A detailed example can be seen in Fig. 5.16, additionally in Fig. 5.19 it is displayed a net of strings and vertices, and at each vertex it is possible to see the triangles made by the incoming framing boxes, as well as the perfect fitting of the various framing polygons just as described in procedure 3.

The volume of the framing box is $|k|^2 = |F(P)| + |F(B(P))|$, the volume of framing polygon is $|k|^2 - |F(B(P))|$. Density inside the framing box is $\rho = 2$, the lost mass in $H(P)$ plus the lost mass in $H(B(P))$ is $|k|^2 - 2$. In $H(P)$ the lost mass is $|F(P)| - 1$.

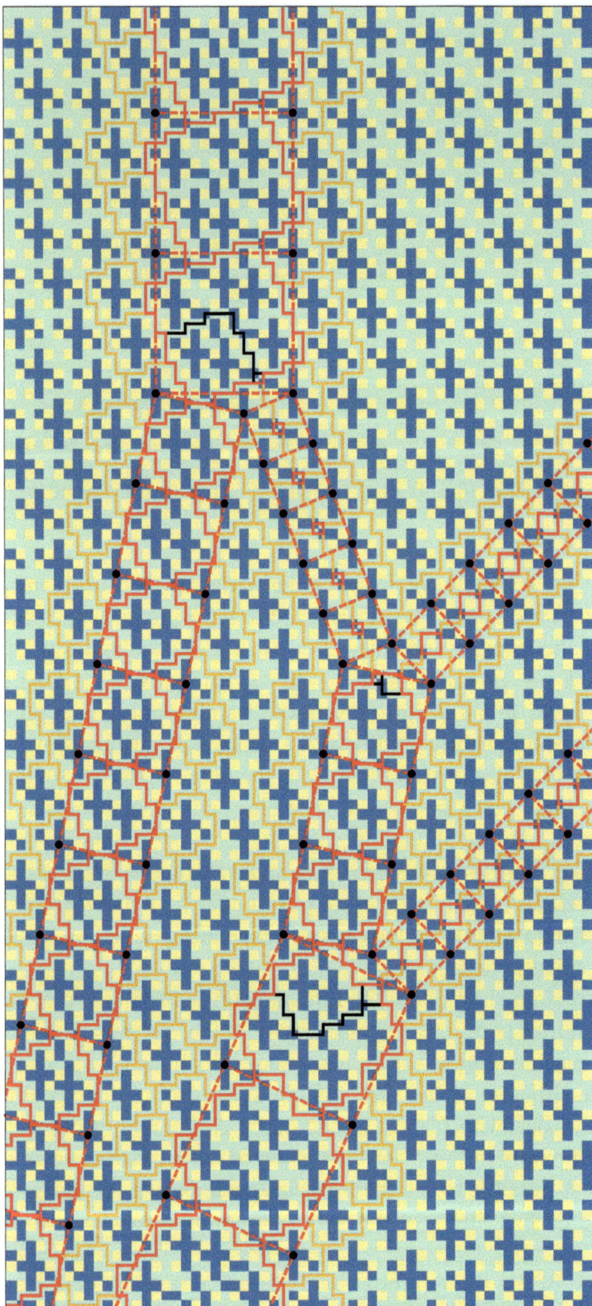

Fig. 5.19 Complex scattering event in the $\Phi(2, -5)$ background, with generators $\boldsymbol{v}_1 = (4, -4)$, $\boldsymbol{v}_2 = (1, 6)$ and $\boldsymbol{v}_3 = -\boldsymbol{v}_1 - \boldsymbol{v}_2 = (-5, -2)$. From *left* to *right* the three vertices are $3\boldsymbol{v}_1 + \boldsymbol{v}_2 \to (2\boldsymbol{v}_1 + \boldsymbol{v}_2) + \boldsymbol{v}_1$ then $2\boldsymbol{v}_1 + \boldsymbol{v}_2 \to (\boldsymbol{v}_1 + \boldsymbol{v}_2) + \boldsymbol{v}_1$ and last $(2\boldsymbol{v}_1 + \boldsymbol{v}_2) + (\boldsymbol{v}_1 + \boldsymbol{v}_2) \to 3\boldsymbol{v}_1 + 2\boldsymbol{v}_2$

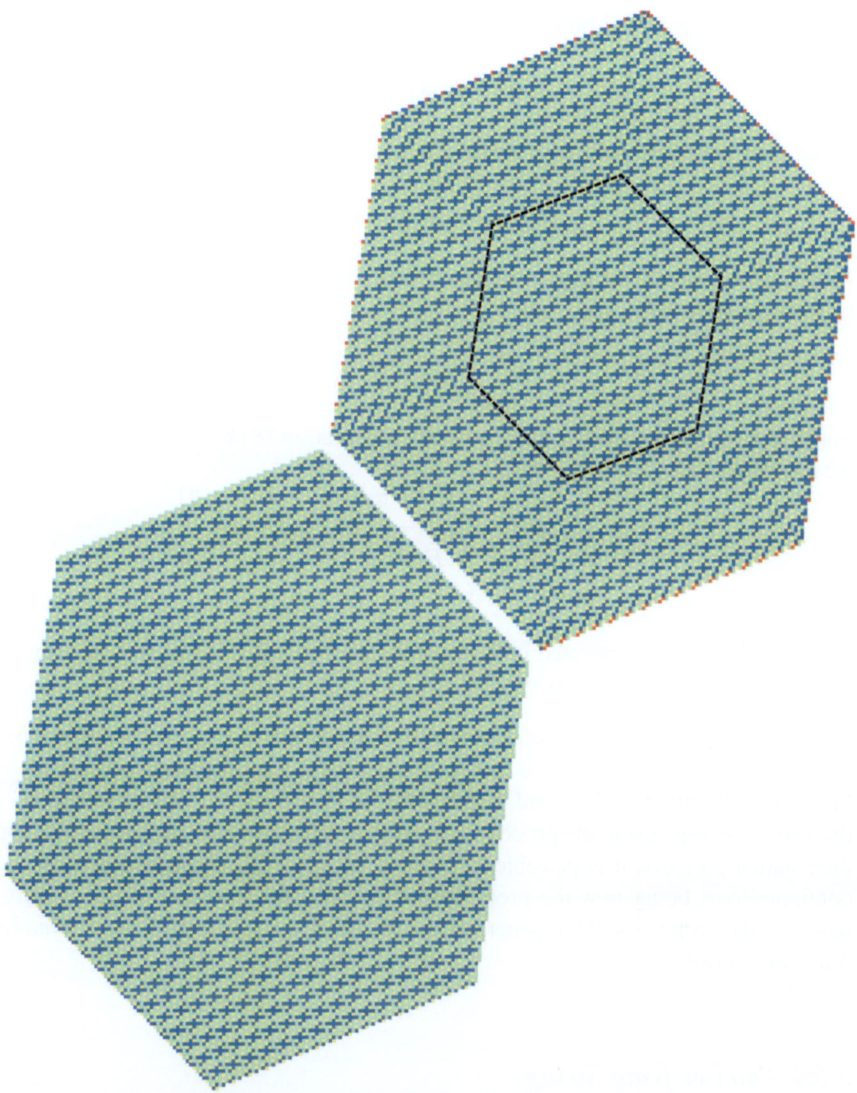

Fig. 5.20 Master protocol with hexagonal open boundaries with the background $\Phi(2, -5)$. The corresponding strings net is composed only by the three type I strings and the fundamental scattering vertices

The density of a framing polygon is $\rho = 2 + \frac{1}{|F(P)|}$ this can be deduced recursively from the following proposition.

Proposition 4. *Given the set* $L_\Sigma(v_1, v_2)$ *of type II strings in background B, where* v_1, v_2 *and* v_3 *are the generators of the background and correspond to the three type I strings; then* $\forall k \in L_\Sigma(v_1, v_2)$ *generated through the procedure 11 the volume of*

its framing polygon is given by

$$|F(k)| = |k|^2 - F(B) \tag{5.21}$$

Proof For the three generators we have that

$$|F(v_\alpha)| = |v_\alpha|^2 - F(B) . \tag{5.22}$$

Procedure 2 generates strings in terms of their component, in doing this it creates a framing polygon $F(k)$ such that

$$F(k) = 2(F(p) + F(q) + F(B)) - F(p - q) . \tag{5.23}$$

Suppose now that the smaller strings satisfy the relation $|F(k)| = |k|^2 - F(B)$ then we can write

$$\begin{aligned}
|F(k)| &= 2(F(p) + F(q) + F(B) - F(p - q) \\
&= 2|p|^2 + 2|q|^2 - F(B) - |p - q|^2 \\
&= |p + q|^2 - F(B) \\
&= |k|^2 - F(B) .
\end{aligned} \tag{5.24}$$

Being all the strings generated in terms of the generators which satisfy the relation (5.22), that is true for every string. $\square$

Now a few words on the construction of the different backgrounds and strings. First of all let stress once more that the background with the smallest unit tile is the maximally filled background, this is why we start from it in the construction of the various strings using the procedure 2. At this point using the function Φ, Ψ and their hatted partners, it is possible to explore the space of all the background and configurations, being now the properties of the "bigger" backgrounds and strings based on the properties of the generators v_1, v_2 and v_3 of $B_{\max}$ which are derived by direct inspection.

5.5.1 Patches from Strings

Given a string P we have a unique (recurrent) background associated to it through the function Φ defined in (5.10). Nevertheless this is not the only possible patch associated to a string.

For example the strings with not coprime momentum components, that can be written as bk for some proper string k and positive integer b, display as a packing of b strings. These packing can be obtained in two different different ways:

1. as a packing of *framing boxes*
2. as a packing of *framing polygons*

in both cases with blue cells at the suitable positions.

In case (1), when the packing is at a framing boxes level (see second image in Fig. 5.8) then the density of the patch is the same of the string, which is $\rho = \rho_{\mathrm{marg}} = 2$. This happens to create a situation in which the burning test works in a sufficiently little patch but it does not if the patch is too big; the limit size is 3, which means that packing a triangular patch with framing box of a string with three tiles per side create a recurrent patch, whether for four or more tiles the patches is not recurrent anymore.

In case (2), when the packing is at a framing polygon's level (see third image in Fig. 5.8) then the density of the patch is less than the density of the string, the precise value being $\rho = \rho_{\mathrm{marg}} - \frac{1}{|F(P)|}$. This happens to create a situation in which the patch is transient whatever size it has.

5.6 The Sierpiński Triangle

The Sierpiński triangle is a structure that arises naturally in many deterministic protocols of the ASM, for example in the patterns covering the identity or the ones growing when adding sand in a single site. It is our aim to prove that the projection on the recurrent space of certain regular configurations (realized as a large triangular patch, with a transient background, surrounded by three suitably chosen recurrent backgrounds) is a fractal of Sierpiński type, i.e. is composed of several different patches, alternately recurrent and transient, organized with the same adjacency structure of a Sierpiński triangle (transient patches are sufficiently small to ensure overall that the configuration is recurrent), and it is exactly the same occurring in the previously seen protocols.

The study of this projection by means of strings, patches and the theory presented in the previous paragraph, will make clear how the Sierpiński triangle is a one shot realization of the whole *zoo* of strings and backgrounds we want to classify.

Let us introduce $\mathcal{T}$, the space of transient periodic configurations. We define a function T that associates a transient tessellation to a given background and an integer n

$$T : \mathcal{P} \times \mathbb{N} \to \mathcal{T}$$
$$(B, n) \to T(B)_n = T_{B,n} \tag{5.25}$$

$T_{B,n}$ is a rectangular configuration composed of four regions, one of which triangular with sides corresponding to the generators of B, covered with different patches.

Three non parallel lines of slope k_1, k_2 and k_3 split the plane in 7 regions, among these regions only one is finite and it is a triangle see Fig. 5.21. We want to inscribe the triangle into a rectangle with sides parallel to the xy axes, the surface of it belong to 4 of the regions splitting the plane, the triangle itself and the three semi-infinite regions lying on the triangle's sides. There can occur a special case when the triangle has vertical or horizontal sides, then the rectangle lies on this side, and its surface

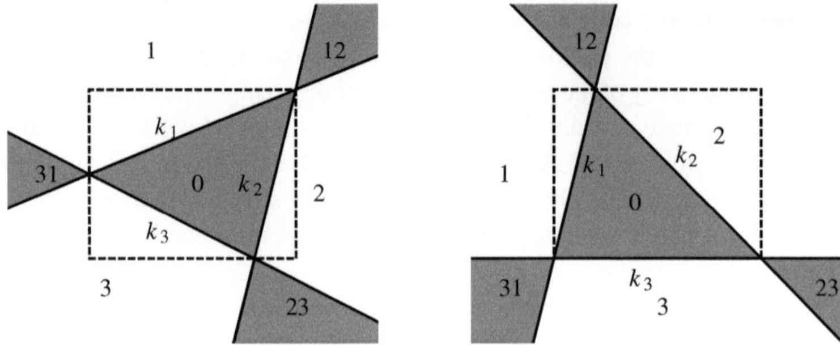

Fig. 5.21 Triangle construction by means of three lines k_1, k_2 and k_3. The *gray* zone 0 is the triangle, then three semi-infinite zones lying on the sides of the triangle are labelled 1, 2 and 3; finally there are three angles 12, 23 and 31. The triangle is inscribed in a rectangle, *dashed line*, covered in general by patches 0, 1, 2 and 3. In the limit case shown on the *right*, just 0, 1 and 2 suffice for the covering

belongs to only three different regions[12].

The configuration $T_{B,n}$ is a rectangle with sides parallel to xy with an inscribed triangle, so we need the filling of only four different regions, in order to completely characterize it.

The triangular region is filled with background tiles with framing polygons $F(B)$ and fillings $H(B)$ arranged along translation vectors k_1, k_2 and k_3 given by

$$k_1 = iv_1, k_2 = iv_2 \text{ and } k_3 = v_3, \tag{5.26}$$

where the v_α are the generators of B; thus each side of the triangle is composed of n tiles and at every point connecting these tiles we place a blue cell. The outer regions 1,2 and 3 in Fig. 5.21, one for each k_α, are filled with the background $\Psi_\alpha(B)$ relative to the string $\hat{\Phi}\alpha(B)$ of momentum k_α. Given the construction, the triangle results in a packing of type II strings, the three $\hat{\Phi}_\alpha(B)$, facing at each side of the triangle the appropriate habitat background $\Psi_\alpha(B)$.

As discussed when introducing $\hat{\Psi}_\alpha$ in (5.11), the limit case of zero size framing polygon can occur for some particular type I strings, it corresponds to a lack of background for the relative type II string $\hat{\Phi}_\alpha(B)$; this absence of background actually means that the side of the triangle sits on the side of the rectangular configuration, with slope k_α for which $|F(\Psi_\alpha(B))| = 0$, and it corresponds to the case when a triangle's side is vertical or horizontal.

The $T_{B,n}$'s are not transient configurations for every value of n. Indeed for $n = 1, 2, 3$ the configuration is recurrent, and it satisfies the burning test[13]. When 4 or more tiles compose the side of the triangle, its interior part do not satisfy the burning test anymore, being $T_{B,n}$ transient for $n \geq 4$. The transient features of

---

[12] Only two regions, if there are two sides, one horizontal and one vertical.

[13] Building the triangle with sides of 1, 2 or 3 tiles, each of the tiles of the triangle is in contact with the exterior, and in this condition the burning test works.

$T_{B,n}$ lies entirely in the triangular region, being the background patches recurrent by construction, in the following we will refer to this triangular region as $Tr_{B,n}$.

Proposition 5. *Given $n \in \mathbb{N}$ and a background $B \in \mathcal{P}$, then $\mathcal{T}(B,n) = T_{B,n}$ is transient for $n \geq 4$ and it is possible to project it on its recurrent representative $[T_{B,n}]$ adding exactly*

$$a(s) = \frac{s(s+1)}{2} \tag{5.27}$$

where $s = \lfloor (n-1)/3 \rfloor$, is called the number of steps.

The nested structure of $[T_{B,n}]$ is Sierpiński like and completely determined in terms of B and n by $\vec{p} = (n_s; p_s, \ldots, p_1) \in \{1,2,3\}^{s+1}$. A summary of the structures is shown in Fig. 5.2

5.6.1 Proof

The mechanism which projects $T_{B,n}$ on its recurrent representative is the key understanding the emergence of the Sierpiński structure. As discussed when we introduced the identity of the sandpile group in Sect. 4.2 the recurrent configuration equivalent to a any given $z \in \mathcal{S}$ can be found by $[z] = Id_r \oplus z$. Being Id_r itself reachable as sum of frame identities Id_f, the same is for the recurrent configuration equivalent to $T_{B,n}$, $[T_{B,n}]$, we are searching for. Hence the projection procedure consists in the addition of a suitable number of frame identities to $T_{B,n}$

$$[T_{B,n}] = T_{B,n} \oplus a(s)Id_f. \tag{5.28}$$

First we note that the addition of a frame identity do not perturb the outer background patches, since these are recurrent, so we need to study in detail the projection mechanism only for the inner triangular region $Tr_{B,n}$.

Let us consider $Tr_{B,n}$ with $n \gg 4$, then the addition of a frame identity transforms its structure as follow: a string $\hat{\Phi}_\alpha(B)$ composed of $n-3$ period is detached at each side of $Tr_{B,n}$ that decrease into $Tr_{B,n-3}$, at the corners of the structure three new triangular regions $Tr_{B'_\alpha,1}$ appear with $B'_\alpha = \Phi(\hat{\Psi}_\alpha(B) + \hat{\Psi}_{\alpha+1}(B))$ at the connection points between three or more different regions are present blue cells, this prescriptions leave some empty space, which is filled by the background B. The addition of a further frame identity detaches an additional string $\hat{\Phi}_\alpha(B)$ from the triangular region, now reduced to $Tr_{B,n-6}$. This string is placed in direct contact with the previous and a string $\hat{\Psi}_\alpha(B) + \hat{\Psi}_{\alpha+1}(B)$ cross the B filled region connecting $\hat{\Phi}_\alpha(B)$ to $\hat{\Phi}_{\alpha+1}(B)$ of the adjacent side. An ulterior frame identity has to be added so that the strings can reach the exterior backgrounds or the $Tr_{B'_\alpha,1}$ respectively. The addition of this two frame identities is a complete *step*. After the second step $Tr_{B,n-3}$ goes into $Tr_{B,n-6}$, two parallel strings $\hat{\Phi}_\alpha(B)$ of $n-6$ periods connect the triangular region to the outer background, the first and the last of this periods have blue cells only in the internal

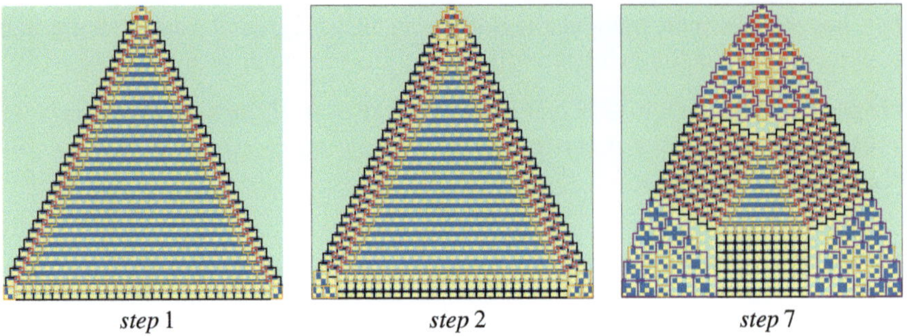

step 1   step 2   step 7

Fig. 5.22 Starting from $Tr_{B,31}$, where B is the background b) in Fig. 5.9, here are shown the first two steps of the projection mechanism, then a further generic step ($s = 7$)) corresponding to the addition of $a(7) = 28$ frame identities

side except the points where they connect with the Tr_B's, and the triangular corner regions grow by one period becoming $Tr_{B'_\alpha,2}$, the remaining space is filled by B. Let stress that, given the duality relation (5.16), when s strings $\hat{\Phi}_\alpha(B)$ with n periods detach from $Tr_{B,n}$ parallel to its side, the situation can be equally described as $n-1$ parallel strings $\hat{\Psi}_\alpha(B)$ of s periods leaving perpendicularly to the side toward the outer background. The generic step s follows an evolution similar to that of the first two steps, we need to add s frame identities to reach step s from $s-1$, and the total number of frame identities to reach it is

$$a(s) = 1 + 2 + \cdots + s = \binom{s+1}{2} = \frac{s(s+1)}{2} \tag{5.29}$$

The result after s steps is composed of three triangular regions $Tr_{B'_\alpha,s}$ at the corners, themselves projected on their recurrent representative $[Tr_{B'_\alpha,s}]$ if $s \geq 4$, a packing of $n - 3s - 1$ strings $\hat{\Psi}_\alpha(B)$ for s periods connecting the outer backgrounds to the inner triangular region, reduced now to $Tr_{B,n-3s}$, and the remaining space is filled with B.

The procedure ends when $Tr_{B,n-3s}$ is recurrent, this correspond to $n' = n-3s \leq 3$ and so we have three cases $n' = 1, 2, 3$. When $n' = 1$, the remaining $Tr_{B,1}$ is actually just a tile of the background B, and being $n - 3s - 1 = n' - 1 = 0$ there are no strings crossing B at all, so the empty space, left when the corner triangular regions $Tr_{B'_\alpha,s}$ are placed, is completely filled by the background B. When $n' = 2$ we are left with a triangle $Tr_{B,2}$ at the center of the triangular structure and the three strings $\hat{\Psi}_\alpha(B)$ departing from it toward the outer backgrounds at each side of $Tr_{B,2}$, the remaining space filled with B. When $n' = 3$ we are left with $Tr_{B,3}$ at the center of the triangular structure and two parallel strings $\hat{\Psi}_\alpha(B)$ departing at each of its sides toward the exterior. The structures with one string $\hat{\Psi}_\alpha(B)$ at each side of $Tr_{B,1}$ are called y and the ones with two parallel string at each side take the name of Y structures, see Fig. 5.2.

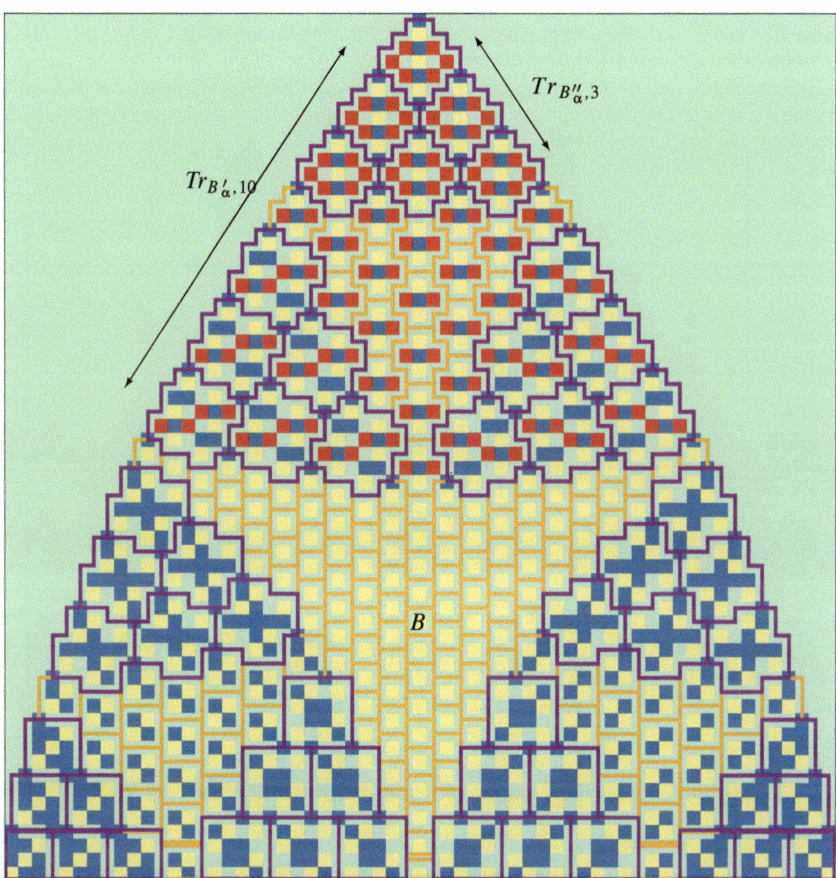

Fig. 5.23 $[T_{B_\alpha,31}]$ being $31 \leftrightarrow (3; 1, 1)$, this Sierpiński has two breaking levels, with no string crossing, and the minimum size triangular unbroken structures have sides composed by three tiles

Given $T_{B,n}$ the number of steps needed to reach the equivalent recurrent configuration $[T_{B,n}]$ depends only on the number of periods and is $s = \lfloor (n-1)/3 \rfloor$, therefore the exact number of frame identities to be added is $a(s) = s(s+1)/2$. The complete nested structure of the Sierpiński can be deduced from the number n of periods for $T_{B,n}$, the recursive procedure starts taking $n = n_0$ then checking step by step if a further breaking level s is possible, $n_{s+1} = \lfloor \frac{n_s-1}{3} \rfloor > 0$ and the interior structure of $[T_{B,n_s}]$ depends on $p_s = n_{s-1} - 3n_s$, $p_s = 1$ corresponds to an imperturbed background B, $p_s = 2$ to y and $p_s = 3$ to Y.

This procedure gives a bijection that associates a vector $\vec{p} = (n_s; p_s, \ldots, p_1)$ in $\{1, 2, 3\}^{s+1}$ to every $[Tr_{B,n}]$ for a suitable s, that corresponds to the breaking level of the Sierpiński; $\vec{p}$ contains explicitly all the information about the inner structure of the fractal. In Table 5.2 the structure of every Sierpiński is given in terms of $\vec{p}$.

Table 5.2 The *periodic table* of geometric Sierpiński structures, for periods $1 \leq n \leq 48$, *dark* and *light blue* regions represent transient and recurrent patches, respectively, while *red lines* correspond to fundamental strings of the corresponding background. For each period number n is given the structure $\vec{p}$. The single **y** structure consists in the intersection of three strings, while the double **Y** structure is the intersection of six

1=(1;)			2=(2;)			3=(3;)		
4=(1;1)	5=(1;2)	6=(1;3)	7=(2;1)	8=(2;2)	9=(2;3)	10=(3;1)	11=(3;2)	12=(3;3)
13=(1;1,1)	14=(1;1,2)	15=(1;1,3)	16=(1;2,1)	17=(1;2,2)	18=(1;2,3)	19=(1;3,1)	20=(1;3,2)	21=(1;3,3)
22=(2;1,1)	23=(2;1,2)	24=(2;1,3)	25=(2;2,1)	26=(2;2,2)	27=(2;2,3)	28=(2;3,1)	29=(2;3,2)	30=(2;3,3)
31=(3;1,1)	32=(3;1,2)	33=(3;1,3)	34=(3;2,1)	35=(3;2,2)	36=(3;2,3)	37=(3;3,1)	38=(3;3,2)	39=(3;3,3)
40=(1;1,1,1)	41=(1;1,1,2)	42=(1;1,1,3)	43=(1;1,2,1)	44=(1;1,2,2)	45=(1;1,2,3)	46=(1;1,3,1)	47=(1;1,3,2)	48=(1;1,3,3)

The geometry of the triangoloid after the projection depends on both B and n (or $\vec{p}$). Indeed after each breaking it is possible to describe exactly the shape of the curve. The breaking level is given through $\vec{p}$ in particular by the value s of its size; at a breaking level s the background B has an interface $B \leftrightarrow Tr_{B'_\alpha}$ that is described by a curve that emerges recursively in the breaking mechanism. The curve has piecewise rational slope given by the corresponding string in background B and has monotone increasing(decreasing) slope. The sequence of slopes in each stretch i of the curve can be written as linear combination of v_α and $v_{\alpha+1}$ the generators of B (slopes of the fundamental strings $\hat{\Psi}_\alpha(B)$ and $\hat{\Psi}_{\alpha+1}(B)$ such that $B'_\alpha = \Phi(\hat{\Psi}_\alpha(B) + \hat{\Psi}_{\alpha+1}(B)))$, so that $k_i = \left(\mathcal{C}_1^s\right)_i v_\alpha + \left(\mathcal{C}_2^s\right)_i v_{\alpha+1}$ and $\left(\vec{\mathcal{C}}^s\right)_i$ defines the curve at each breaking level s. At each level s, i runs from 1 to $2^s - 1$, at the variation of s this is the sequence of Mersenne numbers[14]; given $s \in \mathbb{N}^+$, we obtain $\left(\vec{\mathcal{C}}^s\right)_i$ recursively: we add the auxiliary components $\left(\vec{\mathcal{C}}^s\right)_0 = (1, 0)$ and $\left(\vec{\mathcal{C}}^s\right)_{2^s} = (0, 1)$ to the other components

---

[14] The sequence of Mersenne numbers http://oeis.org/A000225, are of the form $M_n \equiv 2^n - 1$ where n is a positive integer. The Mersenne numbers consist of all 1s in base-2, and are therefore binary repunits. Sometimes Mersenne numbers are considered to be only the ones with n prime.

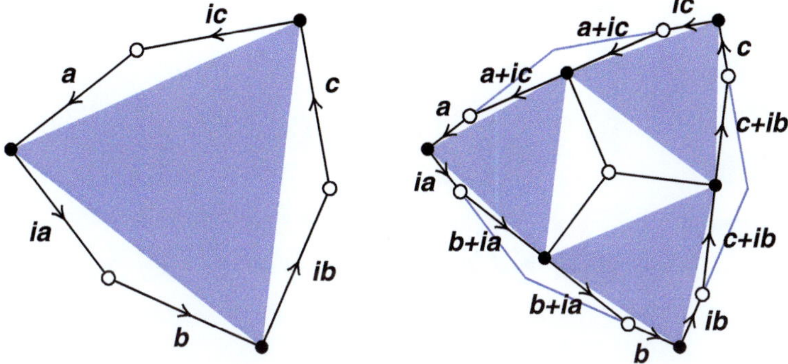

Fig. 5.24 Breaking mechanism from a geometric point of view. Here is evident how the procedure lead to a Bezier curve

Table 5.3 Composition of $\vec{\mathcal{C}}$ in the first three recursive steps

$\vec{\mathcal{C}}^0$:			(1,0)		(0,1)			
$\vec{\mathcal{C}}^1$:		(1,0)		(1,1)		(0,1)		
$\vec{\mathcal{C}}^2$:	(1,0)		(2,1)	(1,1)	(1,2)		(0,1)	
$\vec{\mathcal{C}}^3$:	(1,0)	(3,1)	(2,1)	(3,2)	(1,1)	(2,3)	(1,2)	(1,3)	(0,1)

so $\vec{\mathcal{C}}^0$ has just these two components. Now it is possible to give the recursive step through

$$\left(\vec{\mathcal{C}}^{s+1}\right)_{2i} = \left(\vec{\mathcal{C}}^s\right)_i \quad \text{if } j = 2i \text{ even} \tag{5.30}$$

$$\left(\vec{\mathcal{C}}^{s+1}\right)_{2i-1} = \left(\vec{\mathcal{C}}^s\right)_i + \left(\vec{\mathcal{C}}^s\right)_{i-1} \quad \text{if } j = 2i - 1 \text{ odd.} \tag{5.31}$$

In the Table 5.3 are shown the first three recursive steps where the first and last element of each row correspond to the auxiliary components.

Applying $\vec{\mathcal{C}}$ to the vectors generators of the square lattice $v_1 = (1, 0)$ and $v_2 = (0, 1)$ it is possible generate the set $\mathbb{Z}^2_{\text{gcd}=1}$ of all vectors invariant under $SL(2, \mathbb{Z})$. The set of points $\mathbb{Z}^2_{\text{gcd}=1}$ is called the *Euclid's Orchard*, they are the subset of points of the square lattice visible from the origin. The name Euclid's orchard is derived from the Euclidean algorithm to find the m.c.d. of two numbers. If the orchard is projected relative to the origin onto the plane $x + y = 1$ (or, equivalently, drawn in perspective from a viewpoint at the origin) the tops of the trees form a graph of Thomae's function.

The set of points of the Euclid's Orchard can be reached by a continuous non intersecting curve, which is at each level the one connecting, in order, the points generated through the recursive procedure which gives the exact slope of the Sierpiński triangoloid. We call this curve $\mathcal{E}_\ell$ at each level ℓ and. As shown in Fig. 5.26 the points touched within this procedure are not equidistant from the origin at a given level; nevertheless this curve have an interesting property, not only at each level the curve

Fig. 5.25 The Sierpiński corresponding to the height 2 triangle with a $\pi/2$ angle, inscribed in a *square*. Here are shown in the first row periods $p = 9, 39, 93$ corresponding to $\vec{p} = (1; 1)$, $\vec{p} = (1; 1, 1, 1)$ and $\vec{p} = (3; 1, 1, 1)$, in the second row there is $p = 606$ corresponding to $\vec{p} = (2; 1, 1, 1, 1, 1)$

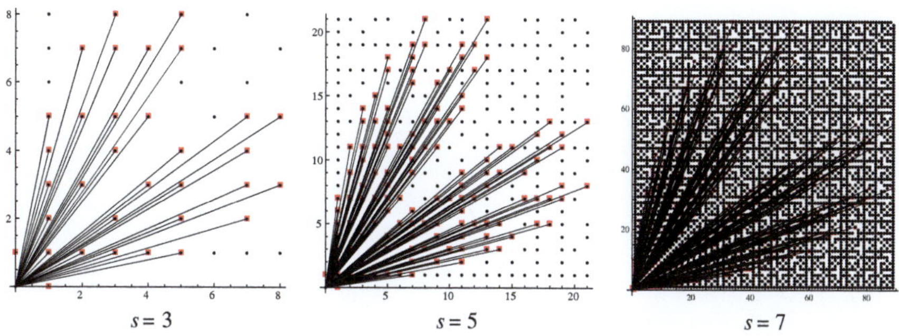

Fig. 5.26 Sequence of $k_i \in \mathbb{Z}^2_{\text{gcd}=1}$ generated by the procedure giving the curve $\mathcal{E}_\ell$ relative to $v_1 = (1, 0)$ and $v_2 = (0, 1)$ at different breaking levels $\ell = 3, 5, 7$

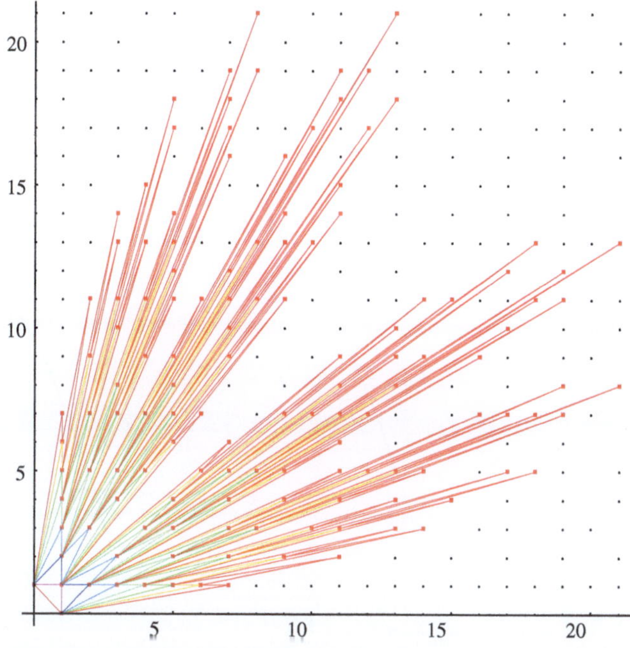

Fig. 5.27 The curves $\mathcal{E}_\ell$ are displayed for different breaking levels ℓ with different colors, at each successive level no intersection appears between the curves

is non intersecting, but also different level curves do not intersect as can be seen in Fig. 5.27

As last remark we note that the convergence of the Sierpiński slope curve is pretty fast as can be seen in Fig. 5.28, and reasoning about the self similarity of the curve generated by the $\vec{C}$ it is possible to confirm that the curve converges in the limit

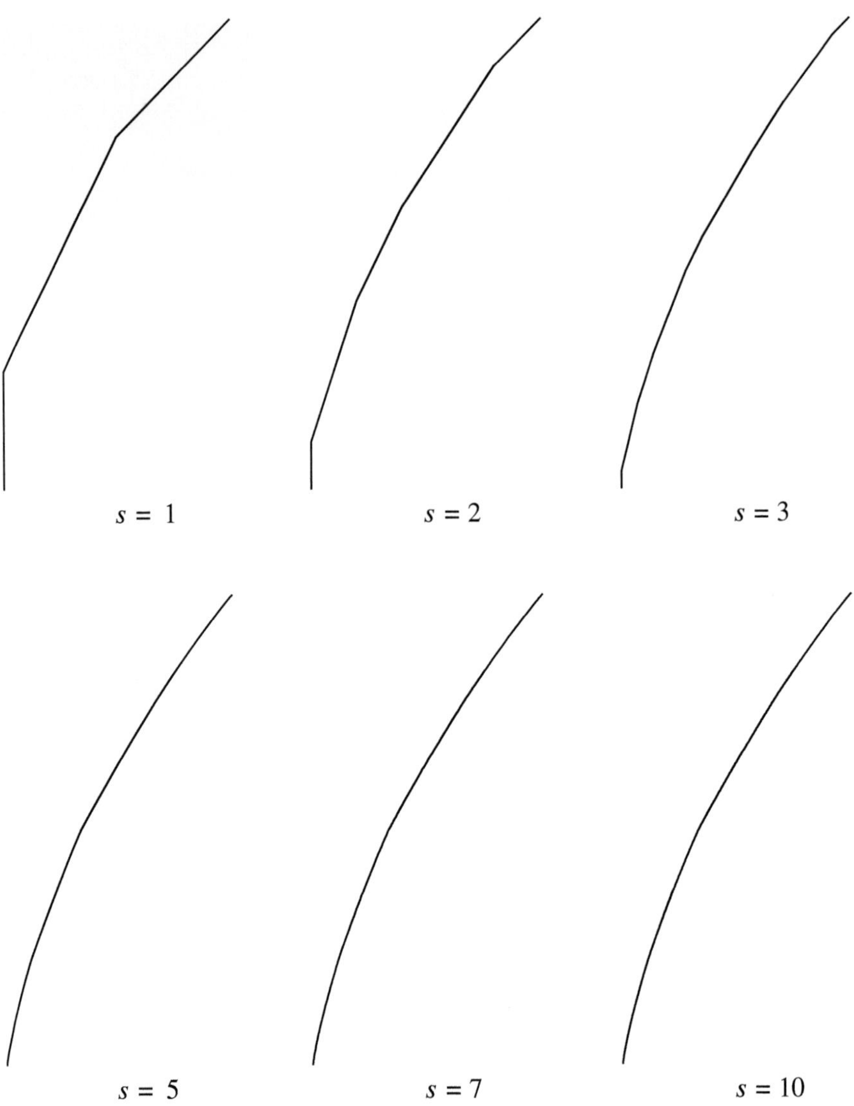

Fig. 5.28 Curves generated by the sequence of k_i relative to $v_1 = (0, 1)$ and $v_2 = (1, 1)$, the maximally filled background, for different breaking levels $s = 1, 2, 3$ and $5, 7, 10$, to show how fast it converges

Fig. 5.29 The Sierpiński in the square lattice relative to bagkground $\Phi(1, 2)$ corresponding to periods $p = 1093$ corresponding to $\vec{p} = (1; 1, 1, 1, 1, 1, 1)$

$\ell \to \infty$ to the generalized Bezier curve for $c = \frac{1}{3}$ whose structure is elucidated in appendix C.

5.6.2 *The Fundamental Sierpiński*

As explained in the case of the duality relation, the maximally filled background $B_{\max}$ together with its string $(1, 1)_{B_{\max}}$ has the particular property (5.18). Similar consideration on the size of the background and string tile that moved us in considering $((1, 1)_{B_{\max}}, B_{\max})$ as the starting configuration when producing the possibles backgrounds and strings suggest to use $B_{\max}$ and study $[T_{B_{\max}, n}]$. Choosing $B_{\max}$ to produce the Sierpiński configuration means to begin with a twice degenerate tri-

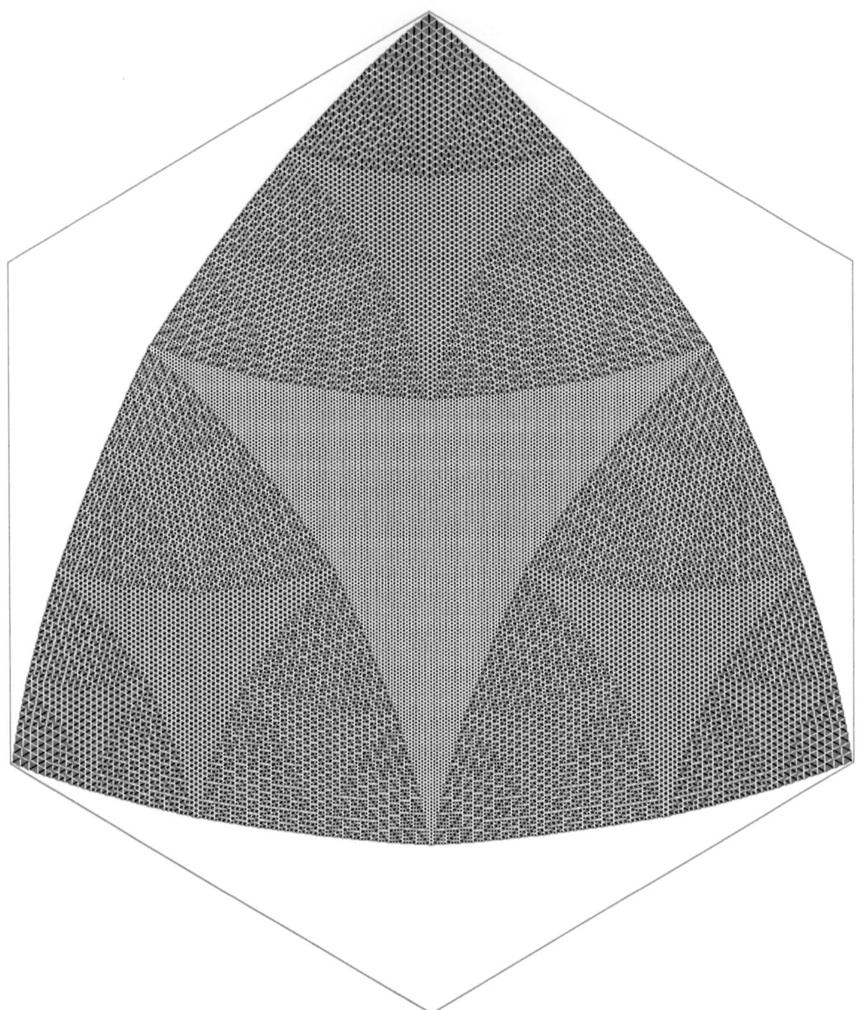

Fig. 5.30 Here we illustrate the triangoloid at the *magic* period number $p = 283$ corresponding to $\vec{p} = (3; 1, 1, 1, 1)$, in the maximally-filled background, as obtained through the protocol introduced in Sect. 5.6. Color code is *white* for $z = 5$, and *darker graytones* for $z = 4, 3, 1$ in this order. Note that, due to the initial height function in the transient patch, and the recursive mechanism of the Sierpiński, some height values (namely $z = 0$ and $z = 2$) do not appear in the configuration

angle, with a $\pi/2$ angle, that is inscribed in a square. The triangle being filled with $(1, 1)_{B_{\max}}$ strings fully packed. In the breaking procedure, one see that while usually the packed configuration till size 3 are recurrent, being transient only for $p \geq 4$, here $p = 3$ is already transient, shifting of one the argument on the relation between the number of periods and the Sierpiński structure. Then the auto-duality play a role in the breaking structure, in such a way that the degenerate triangoloid produces the

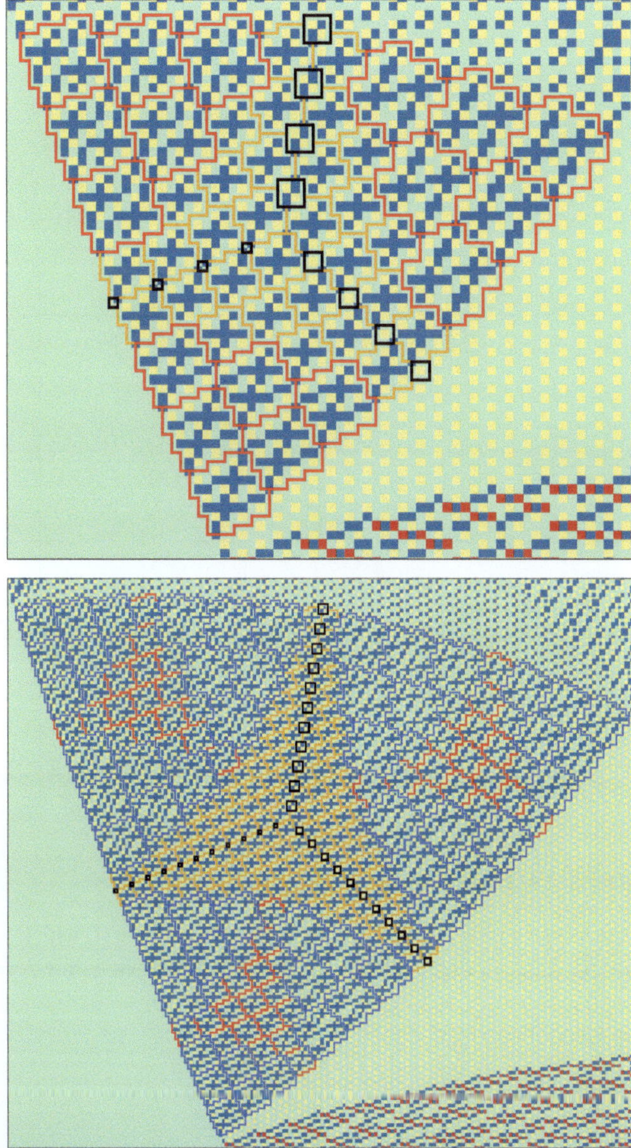

Fig. 5.31 Explanation figures for the Sierpiński nesting. Here are highlighted the tiles for both the recurrent backgrounds and the 'transient' patches. The breaking level is 1 in the top image and 2 in the bottom image

same structure st size $1/3$ in the $\pi/2$ corner. These structures can be seen in Fig. 5.25. If we want to study with more precision the structure of the Sierpiński it is reasonable

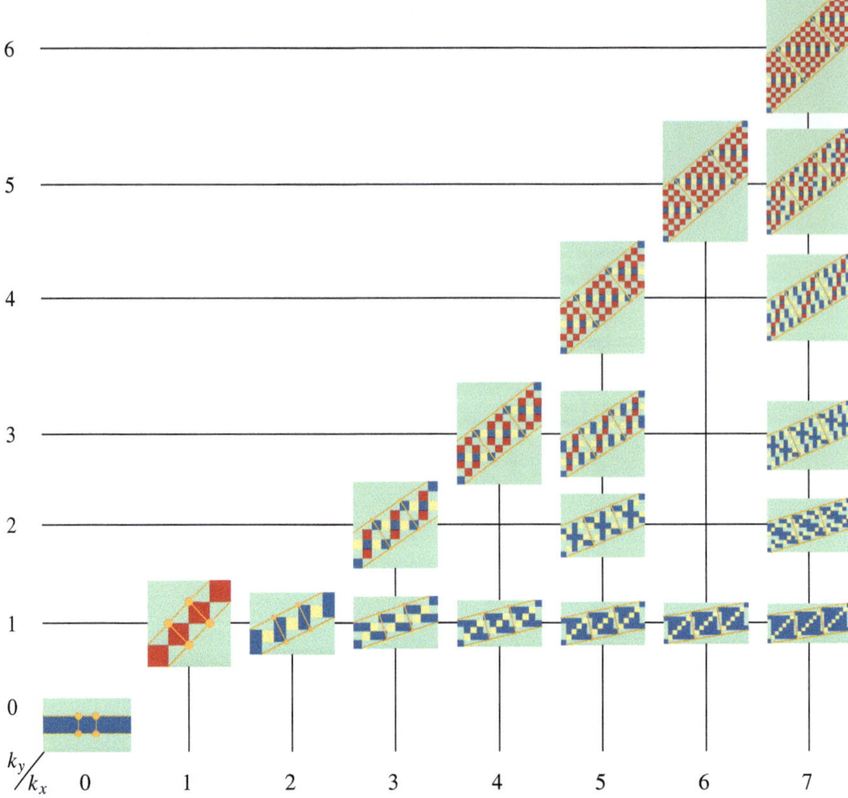

Fig. 5.32 Classification of the strings $\boldsymbol{k} = (k_x, k_y)$ on the maximally filled background. In *yellow* is shown the framing box. Here *yellow circles* sits on the *blue* cells

to focus on the biggest triangoloid composing the fundamental, which results from $T_{B_\alpha, n}$ used in Fig. 5.23. In Fig. 5.29 it is shown such a Sierpiński for $p = 1093$.

5.7 Conclusions

The patterns and patches that emerge naturally in some configurations of the sandpile are attainable under specific protocol in a completely deterministic procedure. This allowed us to study in details their structure discovering the detailed features of these objects.

We describe a framework that shows how *backgrounds* and *strings* are related via a duality relation 10 in such a way that they are both expression of the same framing polygon. This comprehension allows to create a complete catalogue of patches and strings, starting from the maximally filled configuration $z_{\max}$ which corresponds to the maximally filled background, which is the one of minimum size.

In Fig. 5.32 is displayed the family of strings in the maximally filled background up to $|k_x|$ equal to 7, thanks to the particular symmetries of this background these strings suffice to create the whole sett of strings with component of modulus less than 8.

At last is introduced the *Sierpiński* of the sandpile. Its internal structure is completely deduced, not only at a coarse grained level, but the actual value of every single site is given in terms of framing polygons and strings, a few examples of this are shown in Fig. 5.31.

The Sierpiński mechanism of hierarchical classification of patches, as well as the description of strings, hold generically for sandpiles on periodic two-dimensional lattices, that can eventually have different symmetry properties. The triangoloid structure is universal, and, as can be easily inferred, may lead to a dihedrally-symmetric Sierpiński in the case of sandpile on the triangular lattice (thus with heights in the range $0, \ldots, 5 = \bar{z} - 1$). In Fig. 5.25 is shown the fundamental Sierpiński triangoloid for the square lattice, and in Fig. 5.29 the Sierpiński relative to the $\Phi(1, 2)$ background is shown up the breaking level 6; while in Fig. 5.30 is shown the Sierpiński triangoloid in the triangular lattice, in which are evident the symmetries added.

References

1. D. Dhar, T. Sadhu, S. Chandra, Pattern formation in growing sandpiles. Europhys. Lett. **85**, 48002 (2009)
2. T. Sadhu, D. Dhar, Pattern formation in growing sandpiles with multiple sources or sinks. J. Stat. Phys. **138**, 815–837 (2010). doi:10.1007/s10955-009-9901-3
3. D. Dhar, T. Sadhu, *Pattern Formation in Fast-Growing Sandpiles, ArXiv e-prints*, (2011), arXiv:1109.2908v1
4. T. Sadhu, D. Dhar, The effect of noise on patterns formed by growing sandpiles. J. Stat. Mech.: Theor. Exp. **2011**, P03001 (2011). arXiv:1012.4809
5. S. Caracciolo, G. Paoletti, A. Sportiello, Conservation laws for strings in the abelian sandpile model. Europhys. Lett. **90**, 60003 (2010). arXiv:1002.3974v1
6. D. Dhar, Self-organized critical state of sandpile automaton models. Phys. Rev. Lett. **64**, 1613–1616 (1990)
7. P. Bak, C. Tang, K. Wiesenfeld, Self-organized criticality: an explanation of the 1/f noise. Phys. Rev. Lett. **59**, 381–384 (1987)
8. M. Creutz, Abelian sandpiles. Comp. Phys. **5**, 198–203 (1991)
9. Y. Le-Borgne, D. Rossin, On the identity of the sandpile group, Discrete Mathematics. LaCIM 2000 Conf. Combinator. Comp. Sci. Appl. **256**, 775–790 (2002)
10. S. Caracciolo, G. Paoletti, A. Sportiello, Explicit characterization of the identity configuration in an abelian sandpile model. J. Phys. A: Math. Theor. **41**, 495003 (2008). arXiv:0809.3416v2
11. M. Creutz, Abelian sandpiles. Nucl. Phys. B (Proc. Suppl.), **20**, 758–761 (1991)
12. S. Ostojic, Patterns formed by addition of grains to only one site of an abelian sandpile. Phys. A: Stat. Mech. Appl. **318**, 187–199 (2003)
13. A. Fey-den Boer, F. Redig, Limiting shapes for deterministic centrally seeded growth models. J. Stat. Phys. **130**, 579–597 (2008). doi:10.1007/s10955-007-9450-6
14. M. Widom, Bethe ansatz solution of the square-triangle random tiling model. Phys. Rev. Lett. **70**, 2094–2097 (1993)
15. P.A. Kalugin, The square-triangle random-tiling model in the thermodynamic limit. J. Phys. A: Math. General **27**, 3599 (1994)
16. A. Verberkmoes, B. Nienhuis, Triangular trimers on the triangular lattice: an exact solution. Phys. Rev. Lett. **83**, 3986–3989 (1999)

Chapter 6
Conclusions

We want here to give a brief summary of the results reported in this thesis, with particular emphasis on the connections they share and the possible future improvements that they could lead.

As stated in the introduction, the purpose of this work was to study the emergence of patterns in deterministic protocols of the ASM. The first step has been to introduce in Chap. 3 a new algebraic structure in the sandpile. The components of this algebra are the standard operators a_i and the operators $a_i^\dagger$, both acting on the space of stable configurations S. The a_i's add a particle at site i and then, in case of unstable sites, relax the configuration by means of topplings, while the $a_i^\dagger$'s act in a symmetric way, i.e. they delete a particle in i and then anti-relax the configuration in case of sites with negative height. The action of these operators is not commutative, indeed $a_i^\dagger a_i \neq a_i a_i^\dagger$, but when acting on stable configurations it is easy to check they satisfy a Temperley-Lieb like relation $a_i a_i^\dagger a_i = a_i$ and $a_i^\dagger a_i a_i^\dagger = a_i^\dagger$, and in its more general formulation we prove in Theorem 3.7 and its Corollary 3.4 that, for $u \in \Omega$ when acting on $z \in \mathbf{R}$, these operators satisfy

$$a^u (a^\dagger)^u a^u = a^u \tag{6.1}$$

$$(a^\dagger)^u a^u (a^\dagger)^u = (a^\dagger)^u \tag{6.2}$$

Using these operators then we define some projector operators $\Pi_i = a_i^\dagger a_i$ for which a number of relations is proven to be true. These operators when acting on a recurrent configuration can take it into a transient configuration equivalent under toppling.

It is possible to define a Markov Chain on each equivalence class $[z]$ where at each time-step we act with Π_i on a random vertex. Such a Markov chain is in relation to the allowance of multitopplings (defined earlier in Sect. 2.6) in the sandpile dynamics and thus it converges. The final configuration reached is the result of the relaxation of the initial configuration z under the multisite toppling rules, and thus must be such that the maximally filled vertices are isolated. If we choose $z = z_{max}$ as initial configuration, then the final configuration shows the emergence of interesting patterns, in particular

G. Paoletti, *Deterministic Abelian Sandpile Models and Patterns,*
Springer Theses, DOI: 10.1007/978-3-319-01204-9_6,
© Springer International Publishing Switzerland 2014

if we choose a circular geometry it converges to a uniform tessellation of the plane, except some border noise.

Then in Chap. 4 we present a study, numerical and analytical, of the shape of the Creutz identity sandpile configurations, for variants of the ASM, with directed edges on a square lattice (pseudo-Manhattan and Manhattan), and square geometry. Reason for this study was the fact that in this directed graph, the heights are valued in $\{0, 1\}$ (while the original BTW sandpile has heights in $\{0, 1, 2, 3\}$), and we conjectured that this could led to simplifications. We found even simpler results than what we expected, and qualitatively different from the ones in the BTW model.

In the BTW model, the exact configuration seems to be unpredictable: although some general "coarse-grained" triangoloid shapes seem to have a definite large-volume limit, similar in the square geometry and in the one rotated by $\pi/4$, here and there perturbations arise in the configuration, along lines and of a width of order 1 in lattice spacing. We have discussed the role of these structures, both patches and lines of perturbation, in successive chapter.

The triangoloids have precise shapes depending from their position in the geometry, and are smaller and smaller towards the corners. Understanding analytically at least the limit shape (i.e. neglecting all the sub-extensive perturbations of the regular-pattern regions) is a task, at our knowledge, still not completed, although some first important results have been obtained by Ostojic in [1], and further achievements in this direction have come with the work of Levine and Peres [2, 3], both in the similar context of understanding the relaxation of a large pile in a single site

In the Manhattan-like lattices on square geometry, on which we focused, we show how the situation is much simpler, and drastically different. Triangoloids are replaced by exact triangles, all of the same shape (namely, shaped as half-squares), and with straight sides. All the sides of the triangles are a fraction $\frac{1}{2}3^{-k}$ of the side of the lattice (in the limit), where the integer k is a "generation" index depending on how near to a corner we are, and indeed each quadrant of the configuration is self-similar under scaling of a factor $1/3$. The corresponding "infinite-volume" limit configuration is sketched in Fig. 4.9. A restatement of the self-similarity structure, in a language resembling the $z \to 1/z^2$ conformal transformation in Ostojic [1] and Levine and Peres [2], is the fact that, under the map $z \to \ln z$, a quadrant of the identity (centered at the corner) is mapped in a quasi–doubly-periodic structure.

Also the filling numbers (i.e. the minimal number of frame identities relaxing to the recurrent identity) have simple parabolic formulas, while in the original BTW model a parabola is not exact, but only a good fitting formula.

These features reach the extreme consequences in the pseudo-Manhattan lattice, where the exact configuration at given sizes is deterministically obtained, through a ratio-$1/3$ telescopic formula. These work is presented in a paper [4].

Finally in Chap. 5 we describe different experimental protocols producing patterns in the sandpile, and then we describe and explain the different pattern structure in terms of strings and patches. Indeed complex and beautiful patterns can be generated in the ASM and interest on them derive from some peculiar properties they display that are described in the introduction, like allometry. We enumerate a number of deterministic protocols: the generation of the Creutz Identity configuration, the

Markov Chain defined by action of the operators Π_i converging on the Wild Orchids configurations, the Master protocol on the maximally filled configuration or any given allowed background based on the action of the Π_i and finally the construction of the fundamental Sierpiński triangoloid of a given period. On the plane we classify *patches*, or *backgrounds*, patterns periodic in two-dimensions described in term of wallpaper group 2222, and structures periodic in just one dimension, that we call *strings*, and are described in terms of Frieze group 22∞. Strings can be classified in terms of their habitat background and their principal periodic vector k, that we call *momentum*. We derive a simple relation between the momentum of a string and its density of particles, E, which is reminiscent of a dispersion relation, $E = |k|^2$. Strings interact: they can merge and split and within these processes momentum is conserved, $\sum_a k_a = 0$. The role of the modular group $\mathrm{SL}(2, \mathbb{Z})$ is essential behind these laws.

We present a number of functions connection backgrounds with strings, both deriving from and living in the background, this framework allows to identify a duality relation between strings and background proven in Theorem 10 where the action of the duality relation $\mathcal{J}$ has been made explicit. Then we describe in details the construction of the scattering vertices and of higher momentum strings in terms of generating strings as $k \leftarrow (p, q)$ in Sect. 5.5.

Finally we describe in every detail the construction of the Sierpiński triangoloid. In the infinite limit, the Siepiński structure collect all the possible string and patches, ordered in momentum as described by the function $\vec{C}^s$ with s the breaking level of the Siepiński, which is defined through a recursive formula in (5.30). The mechanism of construction of the triangoloid is based on the projection on the space of Recurrent configurations of $T_{B,n}$, which is composed by three recurrent backgrounds and a triangular, transient, patch made by strings derived from B. The projected configuration $[T_{B,n}]$ is built for every number of periods n in terms of patches and strings, site by site.

Of all the possible Siepiński, there exist a fundamental one which collect all the possible strings of the sandpile model. In Fig. 5.29 is shown the one for the square lattice, while in Fig. 5.30 is shown the one for the triangular lattice.

References

1. S. Ostojic, Patterns formed by addition of grains to only one site of an abelian sandpile. Phys. A: Stat. Mech. Its Appl. **318**, 187–199 (2003)
2. L. Levine, Y. Peres, Scaling limits for internal aggregation models with multiple sources. J. Anal. Math. **111**, 151–219 (2010). doi:10.1007/s11854-010-0015-2
3. L. Levine, *Limit Theorems for Internal Aggregation Models*, Ph.D. Thesis, University of California at Berkeley, Fall 2007, arXiv:0712.4358. http://math.berkeley.edu/~levine/levine-thesis.pdf
4. S. Caracciolo, G. Paoletti, A. Sportiello, Explicit characterization of the identity configuration in an abelian sandpile model. J. Phys. A: Math. Theor. **41**, 495003 (2008), arXiv:0809.3416v2

Appendix A
SL(2, $\mathbb{Z}$) a Summary

Consider the following question: given a positive rational number, determine an equivalent expression in the form of a (finite) continued fraction, where only the symbols "1" and "+" are allowed. For example,

$$\frac{25}{9} = 1 + 1 + \cfrac{1}{1 + \cfrac{1}{1+1+1+\frac{1}{1+1}}}$$
$$= 1 + 1 + 1/(1 + 1/(1 + 1 + 1 + 1/(1 + 1))). \tag{A.1}$$

Note that the expression on the second line is redundant: the positions of the symbols "1" and the parentheses are reconstructed univocally from the string of "+" and "/", i.e., by a word in the alphabet $\{+, /\}$, namely in this case

$$w_{25/9} = (+, +, /, +, /, +, +, +, /, +). \tag{A.2}$$

Note that this is a *Fibonacci word*, i.e., in a minimal word, the symbol "/" cannot appear in two consecutive positions (because $1/(1/x) = x$). In order to have more compact notations, we can write k as a shortcut for $1 + 1 + \cdots + 1$ (k summands). Analogously, the Fibonacci word can be equivalently encoded by the length of sequences of consecutive "+". For our example,

$$\frac{25}{9} = 2 + \cfrac{1}{1 + \cfrac{1}{3+\frac{1}{1+1}}}; \qquad \tilde{w}_{25/9} = (2, 1, 3, 1). \tag{A.3}$$

If the number is positive and not rational, the continued fraction, as well as the associated word, are infinite.

Such a structure has some resemblance with the famous decomposition of fractions used in ancient Egypt, and, for brevity, we will call it an *Egyptian continued fraction*.

The procedure to determine the finite continued fraction in the rational case is univocal and elementary, and based on the fact that, calling $\left[\frac{a}{b}\right]$ the continued-fraction

G. Paoletti, *Deterministic Abelian Sandpile Models and Patterns*,
Springer Theses, DOI: 10.1007/978-3-319-01204-9,
© Springer International Publishing Switzerland 2014

expression associated to $\frac{a}{b}$, one has

$$\left[\frac{a}{b}\right] = \begin{cases} 1 + \left[\frac{a-b}{b}\right] & \text{if } a \geq b, \\ \frac{1}{\left[\frac{b}{a}\right]} & \text{if } 0 < a < b. \end{cases} \tag{A.4}$$

The procedure stops when $a = 0$. Note that, at each step, either the denominator decreases, or the numerator decreases, while the denominator remains constant, thus the procedure is guaranteed to stop in a finite number of steps.

Now consider the problem of determining the greatest common divisor of two integers, $\gcd(a, b)$. An algorithm, discussed by Euclid in the *Elements*, is based on the "halting condition" $\gcd(0, b) = b$, and the lemmas $\gcd(a, b) = \gcd(b, a) = \gcd(a + b, b)$, used as a substitutional rule

$$\gcd(a, b) = \begin{cases} \gcd(a - b, b) & \text{if } a \geq b, \\ \gcd(b, a) & \text{if } 0 < a < b. \end{cases} \tag{A.5}$$

The structure is completely analogous to the one shown above for continued fractions. The argument for the fact that the Euclid gcd algorithm halts in a finite number of steps is identical.

Similarly, the shortcut of using integers k instead of strings of $1 + 1 + \cdots + 1$ corresponds, in Euclid algorithm, to replace the largest integer by the remainder of the integer division by the smallest integer (and k is the quotient of the division).

Both Egyptian fractions and Euclid gcd algorithm exist in variants in which the values are in $\mathbb{Z}$, and minus signs are allowed. For example

$$\frac{25}{9} = 1 + 1 + 1 - \frac{1}{1 + 1 + 1 + 1 + \frac{1}{1+1}}. \tag{A.6}$$

Here we used the fact that $\frac{25}{9}$ is "more near to 3 (from below) than to 2 (from above)". The use of minus signs is an improvement in a very special sense: the resulting expression is not guaranteed to use a smaller number of symbols overall w.r.t. the positive case, it is only guaranteed to (possibly) decrease the number of nested fractions. On the Euclid gcd side, this corresponds to (half) the number of steps, under the speed-up of using division remainders.

Define the greatest common divisor for generic integers, by letting $\gcd(a, b) = \gcd(|a|, |b|)$ and $\gcd(a, 0) = |a|$. Thus, gcd is a function from $\mathbb{Z}^2$ to $\mathbb{N}$, and $\mathbb{Z}^2 \setminus 0$ is the disjoint union of the sets $\{\mathbb{Z}^2_{\gcd=k}\}_{k \geq 1}$, preimages of $\mathbb{N}^+$ w.r.t. gcd. The Euclid lemmas $\gcd(a, b) = \gcd(b, a) = \gcd(a + b, b)$ are still valid, and determine symmetry properties of the sets $\mathbb{Z}^2_{\gcd=k}$. A further lemma is $\gcd(ca, cb) = c \gcd(a, b)$, implying that the study of $\mathbb{Z}^2_{\gcd=1}$, a set called *Euclid's orchard*, covers all the cases simultaneously. The two lemmas above (combined with the parity rule $\gcd(a, b) = \gcd(|a|, |b|)$) read

$$(x, y) \in \mathbb{Z}^2_{\gcd=1} \quad \text{iff} \quad \begin{pmatrix} 1 & 1 \\ 0 & 1 \end{pmatrix} (x, y) \in \mathbb{Z}^2_{\gcd=1}; \tag{A.7}$$

$$(x, y) \in \mathbb{Z}^2_{\gcd=1} \quad \text{iff} \quad \begin{pmatrix} 0 & 1 \\ -1 & 0 \end{pmatrix} (x, y) \in \mathbb{Z}^2_{\gcd=1}. \tag{A.8}$$

I.e., the Euclid's orchard is invariant under the action of the matrices $T = \begin{pmatrix} 1 & 1 \\ 0 & 1 \end{pmatrix}$ and $S = \begin{pmatrix} 0 & 1 \\ -1 & 0 \end{pmatrix}$. These matrices have both determinant 1. In fact, they generate the whole group of integer-valued matrices with determinant 1, namely SL$(2, \mathbb{Z})$ (a pictorial representations is given in Figs. A.1, A.2 and A.3).

From the Euclid's Algorithm for the g.c.d. and the egyptian algorithm for the continuous fractions can be deduced an algorithm that decompose the matrices in SL$(2, \mathbb{Z})$ in terms of the generating matrices for SL$(2, \mathbb{Z})$. This procedure is displayed in Sect. A.2.

A.1 Some Simple Properties of SL$(2, \mathbb{Z})$

We start with this elementary fact:

Lemma 1. *Given two positive integers $\vec{k} = (k_x, k_y)$, relatively prime, there exists one and only one quadruplet of non-negative integers $(\vec{p}, \vec{q}) = \big((p_x, p_y), (q_x, q_y)\big)$ such that*

$$\vec{k} = \vec{p} + \vec{q}; \qquad M(\vec{p}, \vec{q}) = \begin{pmatrix} p_x & p_y \\ q_x & q_y \end{pmatrix} \in \text{SL}(2, \mathbb{Z}). \tag{A.9}$$

Define the canonical pair of $\vec{k}$, the pair $(\vec{p}, \vec{q})$ solving the problem above for vector $\vec{k}$.

Proof It is trivially seen that the only canonical pair of $(1, 1)$ is $\big((1, 0), (0, 1)\big)$, and that, if $\mathfrak{r}\big((k_x, k_y)\big) = (k_y, k_x)$, and $(\vec{p}, \vec{q})$ is a canonical pair of $\vec{k}$, a canonical pair of $\mathfrak{r}(\vec{k})$ is $(\mathfrak{r}(\vec{q}), \mathfrak{r}(\vec{p}))$. So we can restrict our attention to $k_x > k_y \geq 1$. It is again trivially seen that the only canonical pair of $(k, 1)$ is $\big((1, 0), (k - 1, 1)\big)$. So we can investigate $k_x > k_y \geq 2$. In this case, it will come out that the entries of $M(\vec{p}, \vec{q})$ are strictly positive.

Indeed, remark that $M(\vec{p}, \vec{q}) \in \text{SL}(2, \mathbb{Z})$ is equivalent to $M(\vec{k}, \vec{q}) \in \text{SL}(2, \mathbb{Z})$, so in particular we must have

$$-q_x k_y \equiv 1 \pmod{k_x}; \tag{A.10}$$

$$q_y k_x \equiv 1 \pmod{k_y}. \tag{A.11}$$

Clearly, for $\vec{k} = \vec{p} + \vec{q}$ to hold with all non-negative entries, we must have $q_x, p_x \leq k_x$ and $q_y, p_y \leq k_y$. As k_x and k_y are relatively prime, the set of values

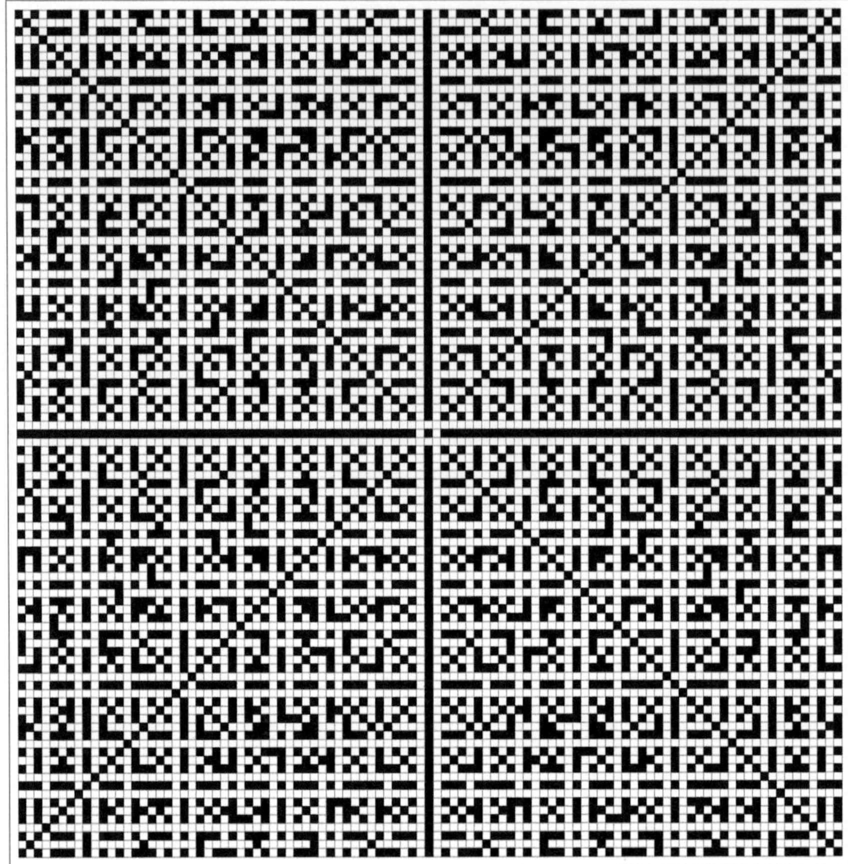

Fig. A.1 The first portion (of side 100) of the set $\widetilde{\mathbb{Z}}_2$, represented as a set of white cells in $\mathbb{Z}_2$

$\{q_y k_x\}_{q_x \in \{1,\dots,k_y-1\}}$ is a permutation of the set $\{1, \dots, k_y - 1\}$, and thus there exists one and only one solution for q_y in the range $\{0, \dots, k_y\}$, and is always attained in the range $\{1, \dots, k_y - 1\}$. An identical reasoning holds for q_x. This completes the proof. $\qquad\square$

A further fact is easily determined:

Lemma 2. *A matrix*

$$M(\vec{p}, \vec{q}) = \begin{pmatrix} p_x & p_y \\ q_x & q_y \end{pmatrix} \in \mathrm{SL}(2, \mathbb{N}), \qquad (A.12)$$

distinct from the identity matrix, has the property that one and only one among $M(\vec{p} - \vec{q}, \vec{q})$ *and* $M(\vec{p}, \vec{q} - \vec{p})$ *is in* $\mathrm{SL}(2, \mathbb{N})$.

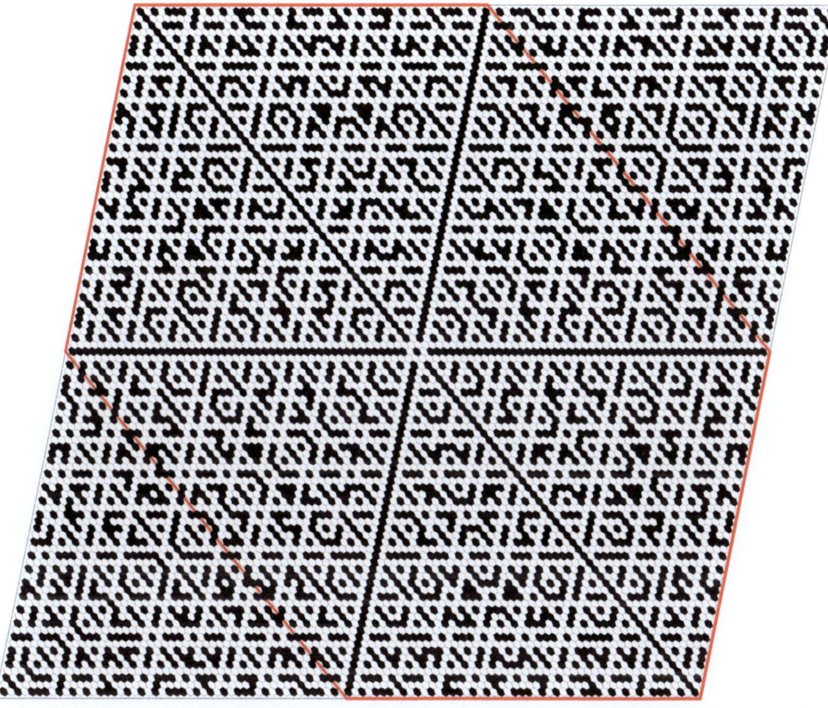

Fig. A.2 The portion of the set $\widetilde{\mathbb{Z}}_2$ as in Fig. A.1, sheared in the x direction, and with cells deformed into *hexagons* accordingly. The presence of six isomorphic sectors is here evident

Proof The fact that the two matrices cannot be both viable is clear, from the fact that $\vec{p} - \vec{q} \neq \vec{0}$, as the vectors are not collinear.

So, what we have to prove is the existence property, i.e. that the fact that one among $\pm(\vec{p} - \vec{q})$ is in $\mathbb{N}^2$, leading to the equation

$$(p_x - q_x)(p_y - q_y) \geq 0. \tag{A.13}$$

Remark that $p_x, q_y \neq 0$, otherwise the determinant would be $-p_y q_x \leq 0$, so we can divide by p_x. Solving the equation $\det M = 1$ w.r.t. q_y gives

$$(p_x - q_x)(p_y - q_y) = \frac{1}{p_x}(p_x - q_x)(p_y p_x - p_y q_x - 1) \tag{A.14}$$

If $p_y = 0$, we have $\frac{q_x - p_x}{p_x}$, but $\det M = 1$ forces $p_x = q_y = 1$, so the only negative case is the identity matrix. Otherwise we can write

$$\frac{1}{p_x p_y}(p_y p_x - p_y q_x)(p_y p_x - p_y q_x - 1).$$

Fig. A.3 (*Color*) The levels in $\tilde{\Omega}$, w.r.t. vectors $(\pm1, 0)$ and $(0, \pm1)$, for integers of absolute value up to 200. *White* cells correspond to pairs $(x, y) \notin \tilde{\Omega}$. For the rest, color code is: *magenta*, *yellow*, *green*, *cyan*, *blue*, *red*, *orange*, for levels from 0 to 6. The directions that asymptotically maximize the growth of the level with the radius are $\pm2^{\pm1/2}$

The product of two consecutive integers is always non-negative. This completes the proof. $\square$

We say that a pair $(\vec{p}, \vec{q})$, distinct from $((1, 0), (0, 1))$, is *of "GS" type* (greater-smaller) if $M(\vec{p} - \vec{q}, \vec{q}) \in \mathrm{SL}(2, \mathbb{N})$, and *of "SG" type* in the other case.

For $\vec{k} \neq (1, 0), (1, 1)$, remark that, if $(\vec{p}, \vec{q})$ is the canonical pair of $\vec{k}$ and it is of GS type, then $(\vec{p} - \vec{q}, \vec{q})$ is the canonical pair of $\vec{p}$, while similarly, if it is of SG type, then $(\vec{p}, \vec{q} - \vec{p})$ is the canonical pair of $\vec{q}$.

A.2 Modular Group Γ

The set of all Möbius transformations of the form

$$\tau' = \frac{a\tau + b}{c\tau + d} \tag{A.15}$$

where a, b, c and d are integers with $ad - bc = 1$, is called the *modular group* and it is denoted by Γ. The group can be represented by 2×2 matrices

$$A = \begin{pmatrix} a & b \\ c & d \end{pmatrix} \quad \text{with } \det A = 1 \tag{A.16}$$

provided we identify each matrix with its negative, since A and $-A$ represent the same transformation. Ordinarily we will make no distinction between the matrix and the transformation. If $A = \begin{pmatrix} a & b \\ c & d \end{pmatrix}$ we write

$$A\tau = \begin{pmatrix} a\tau & b \\ c\tau & d \end{pmatrix}. \tag{A.17}$$

The first theorem shows that Γ is generated by two transformations,

$$T\tau = \tau + 1 \quad \text{and} \quad S\tau = -\frac{1}{\tau} \tag{A.18}$$

Theorem 3. *The modular group Γ is generated by the two matrices*

$$T = \begin{pmatrix} 1 & 1 \\ 0 & 1 \end{pmatrix} \quad \text{and} \quad S = \begin{pmatrix} 0 & -1 \\ 1 & 0 \end{pmatrix} \tag{A.19}$$

That is, every A in Γ can be expressed in the form

$$A = T^{n_1} S T^{n_2} S \cdots T^{n_k} \tag{A.20}$$

where n_i are integers. This representation is not unique.

Proof Consider first a particular example, say

$$A = \begin{pmatrix} 4 & 9 \\ 11 & 25 \end{pmatrix} \tag{A.21}$$

We will express A as a product of powers of S and T. Since $S^2 = 1$, only the first power of S will occur.

Consider the matrix product

$$AT^n = \begin{pmatrix} 4 & 9 \\ 11 & 25 \end{pmatrix} \begin{pmatrix} 1 & n \\ 0 & 1 \end{pmatrix} = \begin{pmatrix} 4 & 4n+9 \\ 11 & 11n+25 \end{pmatrix} \tag{A.22}$$

Note that the first column remains unchanged. By a suitable choice of n we can make $|11n + 25| < 11$. For example, taking $n = -2$ we find $11n + 25 = 3$ and

$$AT^{-2} = \begin{pmatrix} 4 & 1 \\ 11 & 3 \end{pmatrix} \tag{A.23}$$

Thus by multiplying A by a suitable power of T we get a matrix $\begin{pmatrix} a & b \\ c & d \end{pmatrix}$ with $|d| < |c|$. Next, multiply by S on the right:

$$AT^{-2}S = \begin{pmatrix} 4 & 1 \\ 11 & 3 \end{pmatrix} \begin{pmatrix} 0 & -1 \\ 1 & 0 \end{pmatrix} = \begin{pmatrix} 1 & -4 \\ 3 & -11 \end{pmatrix} \tag{A.24}$$

this interchanges the two columns and changes the sign of the second column. Again, multiplication by a suitable power of T gives us a matrix with $|d| < |c|$. In this case we can use either T^4 or T^3. Choosing T^4 we find

$$AT^{-2}ST^4 = \begin{pmatrix} 1 & -4 \\ 3 & -11 \end{pmatrix} \begin{pmatrix} 1 & 4 \\ 0 & 1 \end{pmatrix} = \begin{pmatrix} 1 & 0 \\ 3 & 1 \end{pmatrix} \tag{A.25}$$

Multiplication by S gives

$$AT^{-2}ST^4S = \begin{pmatrix} 0 & -1 \\ 1 & -3 \end{pmatrix} \tag{A.26}$$

Now we multiply by T^3 to get

$$AT^{-2}ST^4ST^3 = \begin{pmatrix} 0 & -1 \\ 1 & -3 \end{pmatrix} \begin{pmatrix} 1 & 3 \\ 0 & 1 \end{pmatrix} = \begin{pmatrix} 0 & -1 \\ 1 & 0 \end{pmatrix} = S \tag{A.27}$$

Solving for A we find

$$A = ST^{-3}ST^{-4}ST^2 \tag{A.28}$$

At each stage there may be more than one power of T that makes $|d| < |c|$ so the process is not unique.

To prove the theorem in general it suffices to consider the matrices $A = \begin{pmatrix} a & b \\ c & d \end{pmatrix}$ in Γ with $c \geq 0$. We use the induction on c.

If $c = 0$ then $a = d = \pm 1$

$$A = \begin{pmatrix} \pm 1 & b \\ 0 & \pm 1 \end{pmatrix} = \begin{pmatrix} 1 & \pm b \\ 0 & 1 \end{pmatrix} = T^{\pm b} \tag{A.29}$$

Thus A is a power of T.

If $c = 1$ then $ad - b = 1$ so $b = ad - 1$ and

$$A = \begin{pmatrix} a & ad - 1 \\ 1 & d \end{pmatrix} = \begin{pmatrix} 1 & a \\ 0 & 1 \end{pmatrix} \begin{pmatrix} 0 & -1 \\ 1 & 0 \end{pmatrix} \begin{pmatrix} 1 & d \\ 0 & 1 \end{pmatrix} = T^a S T^d \tag{A.30}$$

Now we assume the theorem has been proved for all matrices A with lower left element $< c$ for some $c \geq 1$. Since $ad - bc = 1$ we have we have $gcd(c, d) = 1$[1]. Dividing d by c we get

$$d = cq + r, \;\; \text{where } 0 < r < c \tag{A.31}$$

Then

$$AT^{-q} = \begin{pmatrix} a & b \\ c & d \end{pmatrix} \begin{pmatrix} 1 & -q \\ 0 & 1 \end{pmatrix} = \begin{pmatrix} a & -aq + b \\ c & r \end{pmatrix} \tag{A.32}$$

and

$$AT^{-q} S = \begin{pmatrix} a & -aq + b \\ c & r \end{pmatrix} \begin{pmatrix} 0 & -1 \\ 1 & 0 \end{pmatrix} = \begin{pmatrix} -aq + b & -a \\ r & -c \end{pmatrix} \tag{A.33}$$

By the induction hypothesis, the last matrix is a product of power of S and T, so A is too. This complete the proof.

---

[1] If $ad - bc = 1$ then $\gcd(ad, bc) = 1$ but if $\exists g > 1$ s.t. $\gcd(d, c) = g$ then $\gcd(amg, bng) = g$, so $g = 1$.

Appendix B
Complex Notation for Vectors in $\mathbb{R}^2$

In this thesis we deal mostly with systems on regular two-dimensional lattices, and thus makes use of tensor calculus in two dimensions. As well known, the natural isomorphism $\mathbb{R}^2 \simeq \mathbb{C}$ leads to convenient notations, in particular if one has to analyze the behavior of the quantities of interest w.r.t. rotations.

A vector $\mathbf{a} = (a_x, a_y) \in \mathbb{R}^2$ will be identified with a complex number $a = a_x + i a_y \in \mathbb{C}$. Then, the ordinary definitions of scalar and vector product

$$\mathbf{a} \cdot \mathbf{b} := a_x b_x + a_y b_y; \tag{B.1}$$

$$\mathbf{a} \wedge \mathbf{b} := a_x b_y - a_y b_x = \det \begin{pmatrix} a_x & a_y \\ b_x & b_y \end{pmatrix}; \tag{B.2}$$

are rephrased in the complex notation as

$$\bar{a}b = (\mathbf{a} \cdot \mathbf{b}) + \mathbf{i}(\mathbf{a} \wedge \mathbf{b}) \tag{B.3}$$

so that

$$\mathbf{a} \cdot \mathbf{b} = \frac{1}{2}(\bar{a}b + a\bar{b}); \tag{B.4}$$

$$\mathbf{a} \wedge \mathbf{b} = \frac{1}{2i}(\bar{a}b - a\bar{b}). \tag{B.5}$$

Multiplication of vectors in complex notation by a real-positive number corresponds to a dilation, while multiplication by a phase corresponds to a rotation. In particular, multiplying by i corresponds to rotate by $\pi/2$ the vector. As a consequence we have in particular

$$(i\mathbf{a}) \cdot \mathbf{b} = \mathbf{a} \wedge \mathbf{b}; \qquad (i\mathbf{a}) \wedge \mathbf{b} = -\mathbf{a} \cdot \mathbf{b}. \tag{B.6}$$

Already in ordinary "real" notation, it is easy to determine if a polynomial $P(x, y)$ has definite behavior under dilation: $P(\lambda x, \lambda y) = \lambda^k P(x, y)$ if and only if P is homogeneous of degree k. Writing $z = x + iy$, a polynomial $P(z, \bar{z})$ has definite

G. Paoletti, *Deterministic Abelian Sandpile Models and Patterns*,
Springer Theses, DOI: 10.1007/978-3-319-01204-9,
© Springer International Publishing Switzerland 2014

behavior under rotation, $P(e^{i\theta}z, e^{-i\theta}\bar{z}) = e^{i\theta\ell}P(z,\bar{z})$, if and only if P is of the form $P(z,\bar{z}) = z^\ell P(\bar{z}z)$, for $\ell \geq 0$, or of the form $P(z,\bar{z}) = \bar{z}^{-\ell}P(\bar{z}z)$, for $\ell \leq 0$. We call ℓ the *angular momentum* of the quantity. Note that the angular momentum is a scalar in two dimensions.

If a polynomial P has definite behavior both under dilation and rotation, then it is proportional to the monomial $\bar{z}^{\frac{k-\ell}{2}}z^{\frac{k+\ell}{2}}$. In particular, $-k \leq \ell \leq k$, and k and ℓ have the same parity.

A generic homogeneous polynomial of degree k can thus be decomposed as a linear combination of monomials with definite (k, ℓ), for $\ell = -k, -k+2, \ldots, k$. If the polynomial is real-valued, the coefficients of the monomial (k, ℓ) and $(k, -\ell)$ are complex-conjugate, and, if k is even, the coefficient of $(k, 0)$ is real.

In particular, a generic real-symmetric quadratic form

$$\vec{x}^{\mathrm{T}}\omega\vec{x} = \begin{pmatrix} x & y \end{pmatrix} \begin{pmatrix} \omega_{xx} & \omega_{xy} \\ \omega_{xy} & \omega_{yy} \end{pmatrix} \begin{pmatrix} x \\ y \end{pmatrix} \tag{B.7}$$
$$= x^2\omega_{xx} + 2xy\omega_{xy} + y^2\omega_{yy},$$

is decomposed into a part with angular momentum zero (related to the trace of matrix ω), and a part with angular momentum ± 2 (related to the traceless part of ω), i.e., writing (B.7) in terms of $z = x + iy$, one has

$$\vec{x}^{\mathrm{T}}\omega\vec{x} = \frac{1}{4}\left(\omega_2\bar{z}^2 + 2(\mathrm{tr}\,\omega)\bar{z}z + \bar{\omega}_2 z^2\right), \tag{B.8}$$

with

$$\omega_2 = (\omega_{xx} - \omega_{yy}) + 2i\,\omega_{xy}. \tag{B.9}$$

In fact, said tangentially, the customary Pauli matrices

$$\sigma_1 = \begin{pmatrix} 0 & 1 \\ 1 & 0 \end{pmatrix}; \quad \sigma_2 = \begin{pmatrix} 0 & -i \\ i & 0 \end{pmatrix}; \quad \sigma_3 = \begin{pmatrix} 1 & 0 \\ 0 & -1 \end{pmatrix}; \tag{B.10}$$

encode the matrix ω as

$$2\omega = (\mathrm{tr}\,\omega)I + (\omega_{xx} - \omega_{yy})\sigma_3 + 2\omega_{xy}\sigma_1. \tag{B.11}$$

The effect of applying a rotation $R_\theta = \begin{pmatrix} \cos\theta & \sin\theta \\ -\sin\theta & \cos\theta \end{pmatrix}$ to the vector (x, y) corresponds to act on ω with the adjoint, i.e. to transform $\omega \to R_{-\theta}\,\omega\,R_\theta$. The parameter $\mathrm{tr}\,\omega$ remains unchanged (it is a *scalar*) under rotations, while the vector of parameters $(\omega_{xx} - \omega_{yy}, 2\omega_{xy})$ changes with the (left) action of $R_{2\theta}$, i.e., it has angular momentum 2.

Appendix C
Generalized Quadratic Bézier Curve

A *quadratic Bézier curve* is a simple algebraic parametric curve in the plane, determined by three points: the two endpoints (say, at positions x_A and x_B, and a single *control point* (say, at position x_C). Then, the curve is the image of the interval $t \in [0, 1]$, under the map

$$f_{ACB}(t) = (1 - t)^2 x_A + 2t(1 - t) x_C + t^2 x_B. \tag{C.1}$$

This curve (together with the *cubic* variant, having two control points), is widely used in computer vector graphics, because of its peculiar properties: on one side, it is smooth, and its natural parameters encode in a simple way both position and slope at the endpoints (the slopes are along the segments AC and BC)-actually, in the rotated and translated system such that the x-coordinates of (A, C, B) are respectively $(-a, 0, a)$, it is just a parabola; on the other side, it is represented by postscript printers in an efficient way, through a recursive algorithm, based on the following remarkable property:

Proposition 1. *Define*

$$x_{C'} := \tfrac{1}{2} x_A + \tfrac{1}{2} x_C; \tag{C.2a}$$

$$x_{C''} := \tfrac{1}{2} x_B + \tfrac{1}{2} x_C; \tag{C.2b}$$

$$x_D := \tfrac{1}{2} x_{C'} + \tfrac{1}{2} x_{C''}. \tag{C.2c}$$

Then

$$f_{ACB}(t) = \begin{cases} f_{AC'D}(2t) & 0 \le t \le \tfrac{1}{2}; \\ f_{DC''B}(2t - 1) & \tfrac{1}{2} \le t \le 1. \end{cases} \tag{C.3}$$

The proof is by elementary algebra. As a corollary, the support γ_{ACB} of the curve $f_{ACB}(t)$ (i.e., without caring of the "time" parametrization t) is such that

$$\gamma_{ACB} = \gamma_{AC'D} \cup \gamma_{DC''B}, \tag{C.4}$$

G. Paoletti, *Deterministic Abelian Sandpile Models and Patterns*,
Springer Theses, DOI: 10.1007/978-3-319-01204-9,
© Springer International Publishing Switzerland 2014

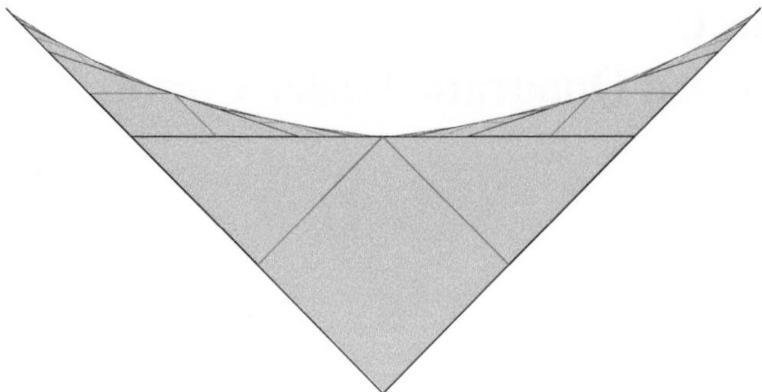

Fig. C.1 The generalized Bézier curve corresponding to our case $c = \frac{1}{3}$, with x_A, x_B and x_C being three of the four vertices of a square. The construction lines of the recursion are shown

and, for any pair of triplets (A, B, C) and (A', B', C'), the curve $\gamma_{A'C'B'}$ is the image of γ_{ACB} under the only affine transformation of the plane mapping (A, B, C) to (A', B', C') (in order). Also, $\gamma_{ACB} = \gamma_{BCA}$ (because $f_{ACB}(t) = f_{BCA}(1 - t)$).

Requiring only these last three properties leads to a generalization of quadratic Bézier curves with a single real parameter $c \in [0, 1]$. Equations (C.2) are generalized to

$$x_{C'} := (1 - c)\, x_A + c\, x_C; \tag{C.5a}$$

$$x_{C''} := (1 - c)\, x_B + c\, x_C; \tag{C.5b}$$

$$x_D := \tfrac{1}{2} x_{C'} + \tfrac{1}{2} x_{C''}. \tag{C.5c}$$

The Bézier case corresponds to $c = \frac{1}{2}$. The geometric construction of Sierpiński profiles, elucidated in Sect. 5.6 and in particular in Proposition 14 corresponds to the case $c = \frac{1}{3}$.

We call the *bisection points of level* k the set of points of the curve which are endpoints of the portions obtained applying the recursive property (C.4), for k steps. We call *bisection points* the union over k of the sets above. This set is dense along the curve (it has roughly the same structure of rational points of the form $a/2^k$ on the unit interval). See Fig. C.1 for a drawing of the curve, that illustrates also this set of points. The bisection points of level k correspond to the incidence points of patches on the boundary of a triangoloid with k breaking levels.

As we explain in a moment, at all values different from $c = \frac{1}{2}$ the curve we obtain is not $\mathcal{C}^\infty$ at all bisection points, where it behaves locally as $f(x) \sim |x|^\gamma$, with $\gamma = \ln \frac{1-c}{2} / \ln c$. Thus the algebraic parametrization of the quadratic Bézier is rather exceptional in this family. In particular, we will also determine that $c = \frac{1}{3}$ is the smallest value of c such that the curve is $\mathcal{C}^1$ (i.e., the first derivative is continuous everywhere), and thus it has a special "threshold" significance.

Given the covariance of the curve under affine transformations, we can equally well assume that $(x_A, x_B, x_C) = ((-1, 1), (1, 1), (0, 1 - 1/c))$, a choice which leads to the simplification $x_D = (0, 0)$, and the curve is symmetric w.r.t. the vertical axis passing through this point.

Let us consider a neighbourhood of x_D. Call $x_B^{(k)}$ and $x_C^{(k)}$ the endpoints and control points of the portion of the curve adjacent to x_D on the right, at the kth level of the recursion. The initial condition is $(x_B^{(0)}, x_C^{(0)}) = ((1, 1), (1 - c, 0))$, and the recursion corresponds to the linear transformation

$$\begin{pmatrix} x_B^{(k+1)} \\ x_C^{(k+1)} \end{pmatrix} = \begin{pmatrix} \frac{1-c}{2} & c \\ 0 & c \end{pmatrix} \begin{pmatrix} x_B^{(k)} \\ x_C^{(k)} \end{pmatrix}. \tag{C.6}$$

For generic values of c, diagonalizing the matrix gives

$$x_B^{(k)} = \begin{pmatrix} \left(1 + \frac{2c(1-c)}{1-3c}\right) \left(\frac{1-c}{2}\right)^k - \frac{2c(1-c)}{1-3c} c^k \\ \left(\frac{1-c}{2}\right)^k \end{pmatrix}. \tag{C.7}$$

Clearly, the sequence of $x_B^{(k)}$ is contained in the curve, and accumulates to x_D. Thus, assuming self-consistently a Lipschitz regularity, we can deduce from (C.7) the behaviour of the curve in a right-neighbourhood of x_D.

A simple analysis for large k (i.e., approaching x_D) gives that, if $c < \frac{1-c}{2}$ (that is, $c < \frac{1}{3}$), the first summand dominates in the x-coordinate, and we have a discontinuity of the first derivative in x_D, by $\frac{2(1-3c)}{1-c-2c^2}$. Conversely, if we have $c > \frac{1}{3}$, the second summand dominates in the x-coordinate, and in a right-neighbourhood of x_D the curve goes like

$$f(x) \sim x^{\frac{\ln \frac{1-c}{2}}{\ln c}}. \tag{C.8}$$

The exponent $\gamma = \ln \frac{1-c}{2} / \ln c$ is a complicated expression, and we expect only sporadic pairs (c, γ) which are 'simple'[2]. One of these pairs is $c = 1/2$, that gives $\gamma = \ln \frac{1}{4} / \ln \frac{1}{2} = 2$, i.e., the expected exact parabolic behaviour.

For the value $c = \frac{1}{3}$ the treatment above is not valid (as can be recognized also from the presence of $1 - 3c$ denominators in (C.7)). What happens is that the matrix in (C.6) is not diagonalizable, because it is a Jordan block $\left(\text{actually, it is just } \frac{1}{3} \begin{pmatrix} 1 & 1 \\ 0 & 1 \end{pmatrix}\right)$. Repeating the reasoning with this explicit matrix gives

$$x_B^{(k)} = 3^{-k} \left(1 + \tfrac{2}{3}k, 1\right) \tag{C.9}$$

from which we get that in a right-neighbourhood of x_D the curve goes like

$$f(x) \sim -\frac{x}{\ln x}. \tag{C.10}$$

---

[2] Although, of course, for generic $\gamma = a/b$ rational, c is algebraic and given by the root of $c^a = \left(\frac{1-c}{2}\right)^b$ in the appropriate range.

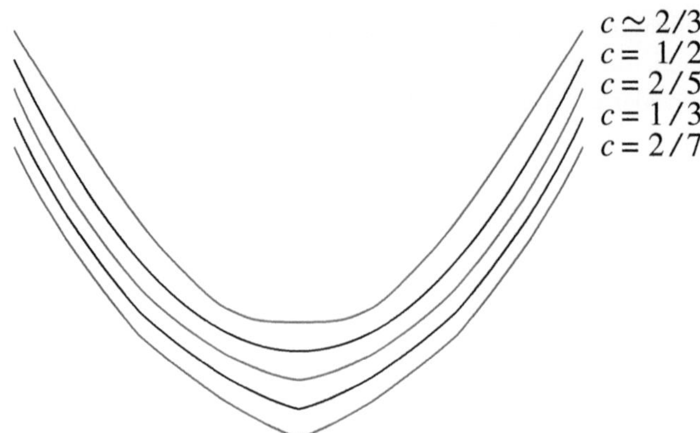

Fig. C.2 The generalized Bézier curve for various values of c. Special values include $c = \frac{1}{2}$ (the ordinary Bézier curve, a parabola), $c = \frac{1}{3}$ (our special case of Sierpiński triangoloids, and also the threshold value at which the curve becomes not $\mathcal{C}^1$), and the curious point $c = 0.6478\ldots$ (root of $2c^4 + c - 1 = 0$, accidentally near to 2/3), for which at all bisection points the curve behaves as $f(x) \sim a_0 + a_1 x + a_4 x^4 + o(x^4)$

In this case the first derivative is still continuous, and the second derivative has an integrable singularity of the form $\sim \frac{1}{x(\ln x)^2}$.

The features of this generalized Bézier curve for different values of c are illustrated through graphical examples in Fig. C.2.

Appendix D
Tessellation

The tessellation, or tiling, of the plane, which is the problem of covering the plane with a given set of elementary tiles, is a subject which emerges spontaneously in nature. Mankind has always been fascinated by it, so that we encounter a number of examples of tessellation in arts. Here we give a brief historical and artistic introduction, then we present the mathematical treatment as given in Grunbaum and Shephard [1]. This is a crucial ingredient for the comprehension of the periodic structures arising in the Abelian Sandpile Model and their mathematical treatment.

D.1 Graphics and Design

Tessellation and tiling emerge in many objects in nature. The honeycomb of honey bees is a spectacular example of a hexagonal tessellation of the space, other tilings of the plane are formed from the breaking lines of a dried mud pond, they can also be recognized in the structures taken by the seeds in a sunflower or the grains of a corn cob on its surface. Finally in the geological formation, such as crystal, some periodic structure in 3-dimensions grow, a cross-section of these structure can be seen as a tessellation of the plane, this is the case at Giant's Causeway in Northern Ireland where basalt columns resulting from an ancient volcanic eruption emerge at the surface resulting in an hexagonal paving of the sea shore.

Since the origin of the civilization, when man started building houses and palaces, he need to cover the floors and the walls placing stones he chose so that the result was agreeable and he was already doing tilings and tessellation in the sense we use these words. Patterns, that is designs repeating motif in a kind of systematic way, must have had a similar origin, as old as that one. Ancient cultures have made large use of tilings. For examples Romans and other Mediterranean people used to produce mosaics to display scenes of life or portraits in their houses; Moors used symmetry to produce the amazing tilings that cover roofs and walls of the chambers in their palaces, examples of such tilings can be seen in *the Alhambra* in Granada or in *the Alcazar* Seville in Spain, see Fig. D.1. Patterns are also present in a huge

G. Paoletti, *Deterministic Abelian Sandpile Models and Patterns*,
Springer Theses, DOI: 10.1007/978-3-319-01204-9,
© Springer International Publishing Switzerland 2014

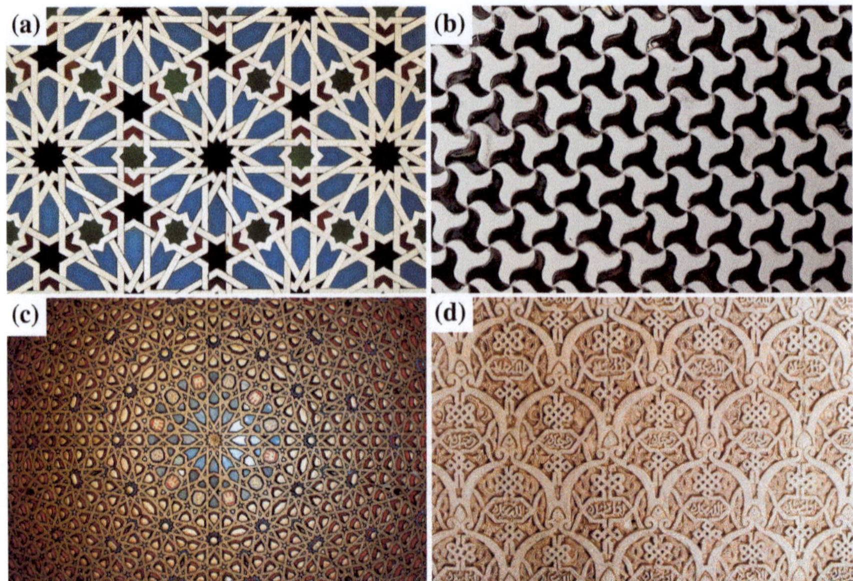

Fig. D.1 Example of several tilings made by moors. **a**, **b**, **c** from the Alcazar and **d** from the Alhambra

number of artifacts through the ages, and many of them, from places distant in time and space, present surprising similarities in the overall composition. This fact has a mathematical reason, that will become clear throughout this appendix.

From many points of view, an extension of the various ideas on the tilings is natural and useful. We shall consider a tiling to be any *partition* of the plane into *regions* regardless whether this partition is realized, or could be, by physical objects. In this sense we understand that there are tilings, or tessellation, all around us, both man-made and in the nature (cells in a membrane, design of a spider-web, the honeycomb of a bee, and so on).

In the past, there have been many attempts to describe, systematize and devise notations for various types of tilings and patterns. However without a mathematical basis such an attempt could not succeed, despite the sometimes prodigious effort devote to them. In the last century a dutch artist Escher (1898–1972), trying to improve his ability in producing interlocking design, and obtaining at first him only primitive results, understood the key role of mathematics in the game. He understood that the types of design he was interested in were governed by groups of symmetries and thus studied the literature available; his results were surprising from an artistic point of view as can be seen for an example in Fig. D.2 and more widely in the collection present in [2]. In his notebooks were found references to some mathematical works by Polya [3], but He also pointed out the importance of a visual approach to the problem, although systematic and with mathematical basis; at this purpose he

Fig. D.2 Example of Escher's tilings

studied and copied the Moorish tile patterns he found in the Alhambra, Granada, Spain. A detailed description of the notebooks can be found in [4].

Tiling and patterns are fascinating subjects. In them visual appeal and ease of understanding combine with possibilities of applying both informally creative and systematic approaches. They concern topics and idea equally useful in art, practical design, crystallographic investigation or mathematical research. Furthermore the *art* of designing tilings is clearly extremely old and well developed. By contrast, the *science* of tilings and patterns, by which we mean the study of their mathematical properties, is comparatively recent and many part of the subject remains unexplored. We present here a mathematical treatment that allows to understand the use of this theory we made in classification of pattern in the Abelian Sandpile Model.

D.2 Tiles, Tilings and Patches

A *plane tiling* $\mathcal{T}$ is a countable family of closed sets $\mathcal{T} = \{T_1, T_2, \ldots\}$ which cover the plane without gaps or overlaps. More explicitly, the union of the sets $T_1, T_2, \ldots$ (which are known as the *tiles* of $\mathcal{T}$) is to be the whole plane, and the interiors of the sets T_i are pairwise disjoint. By "the plane" we mean the familiar Euclidean plane of elementary geometry.

Either of these two conditions can be imposed separately: a family of sets in the plane which has no overlaps is called a *packing*, and a family of sets which covers the plane with no gaps is called a *covering*.

The definition of tiles just given is way too general for our purpose. Indeed it excludes tiles with zero area, but nevertheless it admits tilings in which some tiles have bizarre shape and properties. We restrict our attention on tiles which are a (closed) *topological disk*, by this we mean any set whose boundary is a *single simple closed curve*.

From the definition of tiling we see that the intersection of any finite set of tiles in $\mathcal{T}$ (containing at least two tiles) necessarily has zero area. For most of the tiles considered, this intersection may be empty or a collection of isolated points and arcs.

In these case the point will be called *vertices* and the arcs *edges*. In particular if the tiles are topological disks, then the simple curve which forms the boundary of a tile is divided into a number of parts by the vertices of the tiling, each arc being an *edge of the tile*. Each edge of the tiling coincides with the edges of the two tiles that lie on each side of it. We are interested on tiles with a finite number of vertices. An edge connects two vertices (called the *endpoints* of the edge) and each vertex is the endpoint for a number of vertices. This number is the *valence* of the vertex, and it is at least three. If every vertex of a tiling $\mathcal{T}$ has the same valence j, we say that $\mathcal{T}$ is a *j-valent* tiling. The vertices, edges and tiles of tiling are called its *elements*.

Later we consider the special case in which the tiling is composed by polygons. The usual terminology for polygons has vertices and edges, but in order not to create confusion, we will refer to them as *corners* and *sides* of the polygon. In case the corners and sides coincide of the polygon coincide with vertices and edges of the tiling, we say that the tiling by polygons is *edge-to-edge*. Two tiles are called *adjacent* if they have an edge in common, and then each is called an *adjacent* of the other. Two tiles are called *neighbors* if their intersection is nonempty. Similarly two edges are *adjacent* if they have a common endpoint. The word *incident* is used to denote the relation of a tile to each of its vertices and edges, and also of an edge and each of its endpoints. The relation of *incidence* is considered to be symmetric.

We shall say that two tilings $\mathcal{T}_1$ and $\mathcal{T}_2$ are *congruent* if $\mathcal{T}_1$ may be made to coincide with $\mathcal{T}_2$ by a rigid motion of the plane, possibly including also reflections. Then we say two tilings to be *equal* or *the same* if one of them can be changed in scale (magnified or contracted equally through the plane) so as to be congruent to the other. Equivalently we say that two tiling are equal if there is a similarity transformation of the plane that maps one of the tilings into the other. For example, if $\mathcal{T}_1$ and $\mathcal{T}_2$ are two tilings, each by congruent regular hexagons Fig. D.3(c) then $\mathcal{T}_1$ and $\mathcal{T}_2$ are necessarily equal; for this reason we are justified in referring to *the* tiling by regular hexagons. But $\mathcal{T}_1$ and $\mathcal{T}_2$ are congruent only if the hexagons in $\mathcal{T}_1$ are the same size of those in $\mathcal{T}_2$.

By a *patch* of tiles in a given tiling we mean a finite number of tiles of the tiling with the property that their union is a topological disk—in other words is connected, simply connected, and cannot be disconnected by deletion of a single point.

For the most part of tilings of our interest we shall be concerned with *monohedral* tilings. The word "monohedral" means that every tile in the tiling $\mathcal{T}$ is congruent to one fixed set T, or more simply that all the tiles have the same size and shape. The set T is called the *prototile* of $\mathcal{T}$, and we say that the *prototile* T admits the tiling $\mathcal{T}$. Familiar examples of tiling satisfying all our restrictions are appear in Fig. D.3, these are the *regular* tilings, they are monohedral and their prototiles are regular polygons. Although it seems an easy problem, the study of monohedral tilings is far from being easy, and for example there is not an algorithm to determine whether a given set T is the prototile of a monohedral tiling. The terminology we have introduced can be extended in the obvious way. By a *dihedral* tiling $\mathcal{T}$ we mean one in which every tile T_i is congruent to one or the other of two distinct prototiles T and T'. In a similar way we define *trihedral, 4-hedral, ..., n-hedral* tilings in which there are 3, 4, ..., n

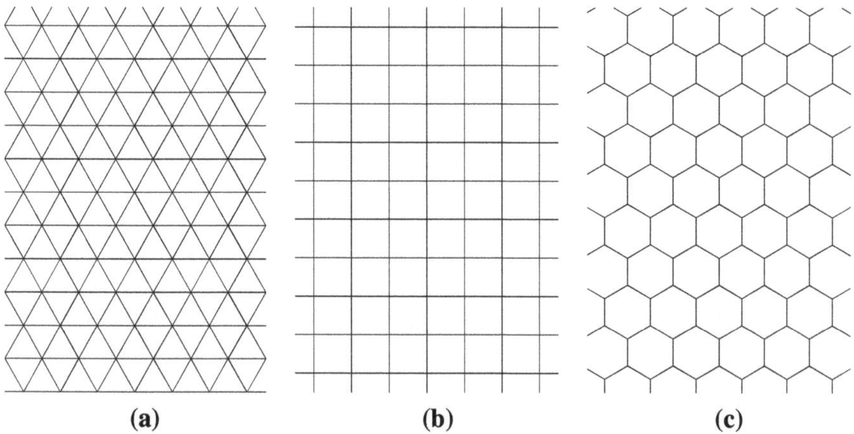

Fig. D.3 The three regular tilings (3^6), (4^4) and (6^3)

distinct prototiles. If the tiling $\mathscr{T}$ uses a set $\mathscr{S}$ of prototiles, we shall say that $\mathscr{S}$ *admits* the tiling $\mathscr{T}$.

D.3 Symmetry, Transitivity and Regularity

Many important properties of tilings depend upon the idea of symmetry. An *isometry* or *congruence transformation* is any mapping of the Euclidean space E^2 in itself which preserves all the distances. Thus if the mapping is denoted by $\sigma : E^2 \to E^2$, and A and B are any two points, the distance between A and B is equal to the distance of their images $\sigma(A)$ and $\sigma(B)$. It is possible to show that very isometry is one of four types:

1. *Rotation* about a opint O through a given angle θ. The point O is called the *center of rotation*. In the particular case when $\theta = \pi$, the line joining A to $\sigma(A)$ will, for all A, be bisected in O, and in this case the mapping is sometimes called *half-turn, central reflection* or *reflection in the point O*.

2. *Translation* in a given direction through a given distance d.

3. *Reflection* in a given line L, the *mirror* or *line of reflection*.

4. *Glide reflection* in which reflection in a line L is combined with a translation through a given distance d parallel to L.

Isometries of type (1) and (2) are usually called *direct* because if points ABC form vertices of a triangle named in a clockwise direction, then the same is true of their images under the isometry σ. If however σ is of type (3) or (4) then the images of the points ABC will form at he vertices of a triangle named in counter clockwise direction. These are called *indirect* or *reflective* isometries.

For any isometry σ and any set S we write σS for the image of S under σ. By a *symmetry* of a set S we mean an isometry σ which maps S onto itself, that is $\sigma S = S$. For example any rotation about the center of a circular disk is a symmetry of the disk, and also the reflections in any line through the center of the disk. In the case of a square, the reflections in the two diagonals and the two axes of the sides are symmetries and so are rotations through angles $\pi/2$, π and $3\pi/2$ about its center O, which is called a *center of 4-fold rotational symmetry*. More in general, if a rotation through $2\pi/n$ around a point O is a symmetry of a given set, then O will be referred as a *center of n-fold rotational symmetry*.

There is an isometry which maps every point onto itself, this is known as the *identity isometry* and is a symmetry of every set. When dealing with rotations we do not distinguish between counterclockwise rotation of θ and clockwise rotation of $2\pi - \theta$, nor between a rotation of θ and a rotation of $\theta + 2k\pi$, for any integer k. As symmetries they are regarded as identical, this because only the final result of the mapping does matter, not the means of arriving at the result.

For any T we denote by $S(T)$ the set of symmetries of T. This set has algebraic properties, the symmetries can be combined by applying them consecutively and the result is another symmetry. Because of this algebraic property $S(T)$ is known as a group, and the number of symmetries in $S(T)$ is called the *order* of the group.

It is convenient to introduce some notation for groups that occur frequently. We shall use $c1$ or e for the group with only one isometry (the identity), and cn ($n \geq 2$) for the group consisting of rotations through angle $2\pi j/n$ ($j = 0, 1, \ldots, n-1$) about a fixed point. This is called the *cyclic group of order n* and is the symmetry group of the "n-harmed swastika". Finally, we use dn ($n \geq 1$) for the group which include all the symmetries of cn together with reflections in n lines equally inclined to one another. This is called the *dihedral group of order 2n*; for $n \geq 3$ is the symmetry group of the regular n-gon. When $n = 1$, the group $d1$ (of order 2) consists of just the identity and the reflection on a line; when $n = 2$, the group $d2$ (of order 4) consists of the identity, reflections on two perpendicular lines and rotation through angle π around the point in which the two lines of reflection meet. The *rotation group $d\infty$* consists of all rotations about a point and all the reflections in lines through that point; it is the symmetry group of a circular disk. Note that the groups cn and dn ($n \geq 1$) each have a property of leaving at least one point of the plane fixed; in fact these are only the groups that can occur as symmetry groups $S(T)$ of *compact* (that is closed and bounded) sets T.

We extend the definition of symmetry in a natural way to structures more complicated than single sets. Thus in the case of a tiling $\mathcal{T}$ we say that an isometry σ is a *symmetry of $\mathcal{T}$* if it maps every tile of $\mathcal{T}$ onto a tile of $\mathcal{T}$. An easy and informal way to think of a symmetry of a tiling is the following. Imagine we have drawn the tiling on an infinite piece of paper, and hen traced it onto a transparent sheet. A symmetry corresponds to a motion of the latter (including the possibility to turn it over) such that, after the motion, the tracing fits exactly over the original drawing. The idea of a symmetry can be extended to more general situations. Suppose, for example, we have a *marked tiling*, one in which there is a *marking motif* on each tile. Then a symmetry of the marked tiling is an isometry which not only maps the tiles of $\mathcal{T}$

onto tiles of $\mathcal{T}$, but also maps each marking on a tile of $\mathcal{T}$ onto a marking on the image tile. Thus in the informal interpretation above, we do not only trace tiles, we trace marks too. The same argument apply for a *colored tiling*.

For any tiling $\mathcal{T}$, we extend the notation introduced above and write $S(\mathcal{T})$ for the group of symmetries of $\mathcal{T}$. It is, of course, possible for $S(\mathcal{T})$ to be the identity alone or it may have many symmetries. This facts can be used as a basis classification, we will discuss this in more detail in Sect. D.4.

If a tiling admits any symmetry in addition to the identity symmetry then it will be called *symmetric*. If its symmetry group contains at least two translations in non-parallel directions then the tiling will be called periodic. many of the tilings we shall meet are periodic and they will be very easily described. Let us represent the two non-parallel translations by vectors a, b. Then clearly $S(\mathcal{T})$ contains all the translations $na + mb$ where n and m are integers. All these translations arise by combining n of the translations a and m of the translations b. Starting from any fixed point O the set of images of O under the translations $na + mb$ forms a *lattice*. The most familiar example of lattice is the set of points in the Euclidean plane with integer coordinates. This is known as the *unit square lattice*, see Fig. D.4; we have already met it as the set of vertices of the regular tiling (4^4) in Fig. D.3. More generally, a lattice can be regarded as consisting of the vertices of a *parallelogram tiling*. Thus with every periodic tiling $\mathcal{T}$ is associated a lattice, and the points of the lattice can be regarded as the vertices of a parallelogram tiling $\mathcal{P}$ see Fig. D.4; the tiles of $\mathcal{P}$ are known as *period parallelograms*. If we know the configuration formed by the tiles, edges and vertices of $\mathcal{T}$ that are contained in one of the parallelograms of $\mathcal{P}$, then the rest of $\mathcal{T}$ can be constructed by repeating this configuration in every parallelogram of $\mathcal{P}$. Not periodic tiling are possible, for example tilings with just one translation vector, or tilings with only rotational symmetry.

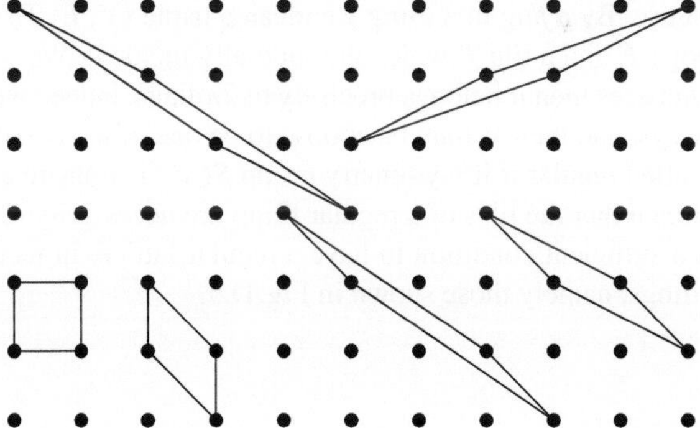

Fig. D.4 A lattice Λ of points in the plane and some *parallelograms* whose corners coincide with points of Λ. Each of these *parallelograms* is the prototile of the corresponding tiling whose vertices coincide with Λ. In fact any *parallelogram* may be chosen as long as the only points it contains are its vertices, and no lattice point lies in its interior or on its boundary. All such *parallelograms* have equal area

Let T be a tile of the tiling $\mathscr{T}$. Then every symmetry of $\mathscr{T}$ which maps T onto itself is clearly a symmetry of T. But the converse is not true, in general. Hence we must distinguish between $S(T)$, the group of symmetries of the tile T, and $S(\mathscr{T}|T)$, the group of symmetries of the tile T which are also symmetries of the tiling $\mathscr{T}$. For brevity we shall often refer we shall often refer to $S(\mathscr{T}|T)$ as the *induced tile group* or as the *stabilizer* of T in $\mathscr{T}$.

Two tiles T_1 and T_2 of a tiling $\mathscr{T}$ are said to be *equivalent* in the symmetry group $S(\mathscr{T})$ contains a transformation that maps T_1 onto T_2; the collection of all tiles of $\mathscr{T}$ that are equivalent to T_1 is called the *transitivity class* of T_1. If all tiles of $\mathscr{T}$ form one transitivity class we say that $\mathscr{T}$ is *tile-transitive* or *isohedral*, the regular tilings are isohedral. The distinction between isohedral tilings and monohedral tilings (in which each tile has the same shape) may seem slight, but it is very significant.[3] If $\mathscr{T}$ is a tiling with exactly k transitivity classes then $\mathscr{T}$ is called *k-isohedral*. Generally, of course, if the tiles are of n different shapes then there will be at least n transitivity classes. In the case of a non symmetric tiling, every tile is a transitivity class on its own.

The idea of transitivity and equivalence is applicable to other elements of a tiling also. If the symmetry group $S(\mathscr{T})$ of $\mathscr{T}$ contains operations that map every vertex of $\mathscr{T}$ onto any other vertex, then we say that the vertices form one transitivity class, or that the tiling is *isogonal*. In an analogous manner to that defined above, we may say that a tiling is *k-isogonal* if its vertices forma k transitivity classes, where $k \geq 1$ is any integer.

A *monogonal* tiling is one in which every vertex, together with its incident edges, forms a figure congruent to that of any other vertex and its incident edges. The distinction between isogonal and monogonal tilings is analogous to that between isohedral and monohedral tilings. *Isotoxal* tilings are tilings in which every edge can be mapped onto any other edge by a symmetry of he tiling.

In order to define a regular tiling we again use the concept of transitivity, but in a very strong sense. By a *flag* in a tiling we mean a triple (V, E, T) consisting of a vertex V, an edge E and a tile T which are mutually incident. We see that if T has n edges and n vertices then it belongs precisely to $2n$ flags, indeed we can choose E in n different ways, and then V may be choose to be one of the two endpoints of E. A tiling $\mathscr{T}$ is called regular if its symmetry group $S(\mathscr{T})$ is transitive on the flags of $\mathscr{T}$. It can be shown that the tiles of a regular tiling are necessarily regular polygons but, this is not a sufficient condition to have a regular lattice. In fact there are only three regular tilings, namely those shown in Fig. D.3.

---

[3] i.e. the problem of classifying the monohedral tilings is unsolved, whereas it is possible to classify all the isohedral tilings.

Table D.1 Symbols used in the representations of the symmetry in the diagrams

Symbol	Meaning
———	Line of reflection
··········	Line of a glide-reflection.
◊	Center of a 2-fold rotation (reflection on a point).
△	Center of a 3-fold rotation.
□	Center of a 4-fold rotation.
○	Center of a 6-fold rotation.
○——▸	Vector indicating the translation in the group.

D.4 Symmetry Groups of Tiling: Strip Group and Wallpaper Group

Here we call "elements" the reflections, glide-reflections, rotations and translations in $S(\mathcal{T})$. In Table D.1 we explain in detail the symbol we use to represent the elements diagrammatically.

A diagram in which the elements have been represented in this manner serves to define the group $S(\mathcal{T})$ precisely. we shall call it a *group-diagram* for $S(\mathcal{T})$.
The representation of the translations by arrows differs from that of the other elements in two important respects:

1. Only the magnitude and the direction of the arrow (vector) is important; unlike the lines of reflection and glide-reflection and centers of rotation, its actual position relative to the tiling is irrelevant. Thus it may be moved parallel to itself and it will still represent the same translation. Technically it is known as a *free vector*, that is, one that does not act at a particular fixed point or along a fixed line.

2. If $S(\mathcal{T})$ contains a translation, then it must contain an infinity of such. However, on the group diagram it is only necessary to indicate at most two. If the translations in $S(\mathcal{T})$ are all parallel then they can be represented as the set of vectors $\{na\}$ where a is a fixed vector and n runs through the integers, positive negative and zero. Hence we need to use only one arrow (representing a or $-a$) to specify all the translations in $S(\mathcal{T})$. If, on the other hand, $S(\mathcal{T})$ contains non-parallel translations, then the tiling is periodic and the set of all the translations may be written as $\{na+mb\}$, where a and b are fixed vectors and n, m run independently through the integers. Thus the translations in $S(\mathcal{T})$ can be specified by just two arrows, one corresponding to a and one to b. However it should be observed that, as we have seen, the choice pf a and b is not unique. (The vectors a, b corresponding to any two adjacent sides of the parallelogram in Fig. D.4 yield the same lattice and therefore the same set of translations.)

Because of these properties it is convenient to regard two group diagrams as the same if they can be made identical by movement (rigid motion) or by altering one's choice of vectors corresponding to translations as described above. In other words, we shall not distinguish between diagrams corresponding to the same group—or between group diagrams that differ trivially.

In classifying symmetry groups, the most important concept is that of *isomorphism*. We say that two symmetry groups $S(\mathscr{T}_1)$ and $S(\mathscr{T}_2)$ are *isomorphic* if the group diagram of one can be made to coincide with the group diagram of the other by applying a suitable affinity. An *affinity* (or *affine transformation*) is defined as any mapping of the plane onto itself representable by linear equations of the form

$$x' = px + qy + c$$
$$y' = rx + sy + d, \tag{D.1}$$

with $ps - qr \neq 0$. Geometrically any lattice can be brought into coincidence with any other lattice by applying a suitable affinity. Thus an affinity can be described as built up by successively applying a rigid motion, a change of scale, and a shear (that is a change of angles between axes).

If $S(\mathscr{T}_1)$ and $S(\mathscr{T}_2)$ are isomorphic, then we shall say that $\mathscr{T}_1$ and $\mathscr{T}_2$ are of the same *symmetry type*. It is perhaps surprising that although the number of symmetry groups $S(\mathscr{T})$ is clearly infinite, if we restrict attention to those tilings whose tile are topological disks, the number of symmetry types is very limited. In fact when $S(\mathscr{T})$ contains no translations, it must be one of the types cn or dn defined in Sect. D.3, and if it does contain translations then it must be one of 24 types. In Figs. D.5, D.6, D.7 and D.8 we display the group diagrams of all these types. We remark here a useful fact if one is trying to evaluate symmetries empirically: *every symmetry in $S(\mathscr{T})$ is also a symmetry of the group diagram of $S(\mathscr{T})$*. Thus, for example, every group diagram is necessarily symmetric with respect of every line of reflection it contains. If we start with a line of reflection and a center of n-fold rotation not lying on it, then using this principle we can generate a large part, or possibly all, of the group diagram. The enumeration problem can therefore be solved by carrying out this procedure systematically, starting by various group elements and doing in a way that does not miss any possibility. Let point out that a group diagram possess more symmetries than the tiling from which it originated.

It must be emphasized that this analysis and claim for completeness only hold when the tiles are topological disks, otherwise many other symmetry types are possible. a detailed derivation of the 7 frieze groups and the 17 crystallographic groups can be found in [5]. There you can find also a key to determine the type of symmetry group of a given tiling or pattern, other keys can be found in [6, 7].

Now let us consider the groups $S(\mathscr{T})$ in more detail. It is convenient to begin considering three cases distinguished by the existence or other wise of translations in $S(\mathscr{T})$:

1. *$S(\mathscr{T})$ contains no translations.* The non-trivial symmetries are necessarily rotations and reflection and not glide-reflections. One possibility is that $S(\mathscr{T})$ con-

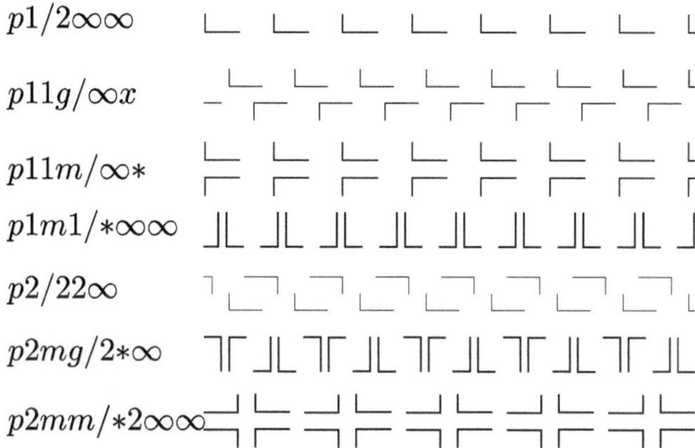

Fig. D.5 The seven frieze groups. Symbols are given in *Crystallographic/Orbifold* notation

tains rotations only, and then is of the type *cn*, the cyclic group of order *n*, for some value of *n*. If it contains more than one reflection then the corresponding lines of reflections cannot be parallel (for otherwise $S(\mathscr{T})$ would contain a translation) and therefore they meet in a point *P*. The product of the two reflections will be a rotation around *P* through an angle 2*v*, where *v* is tha angle between the lines of reflection. The fact that the tiles are topological disks implies that *v* must be a rational multiple of π. Hence $S(\mathscr{T})$ is of finite order, it is of type *dn*, the dihedral group of order 2*n*, for some value of *n*. Since we conventionally use *c*1 and *d*1 for the identity group and the group containing only one reflection, we can assert that any symmetry group $S(\mathscr{T})$ without translations is of type *cn* or *dn* with $n \geq 1$. In any case it is noted that there is at least one point of the plane (the center of the tiling) that is left fixed by every symmetry of $\mathscr{T}$.

2. $S(\mathscr{T})$ *contains translations, all of which are parallel to a given direction L.* Here $S(\mathscr{T})$ can contain reflections in lines perpendicular to *L*, and also at most one reflection in a line parallel to *L*. the only rotations that can occur are 2-fold, through angle π, and, if these exist, their centers must lie at equal distances on a line parallel to *L*. Only seven different types of groups like this can arise, they are shown in Fig. D.5. The seven groups are denoted by *p*1, *p*11*g*, *p*11*m*, *p*1*m*1, *p*2, *p*2*mg* and *p*2*mm* using the notation explained in Sect. D.4.1. They also occur as symmetry group of strip patterns and for this reason they are called *strip groups* or *frieze groups*. Any symmetry of the resulting plane tiling must superimpose the strip onto itself.

3. $S(\mathscr{T})$ *contains translations in non parallel directions.* Here the tiling $\mathscr{T}$ is periodic and $S(\mathscr{T})$ can contain reflections, glide-reflections and rotations. It turns out the the only rotations that can occur are of orders 2, 3, 4 or 6 and that there are only 17 distinct types of groups in all. These are indicates in Figs. D.6, D.7 and D.8.

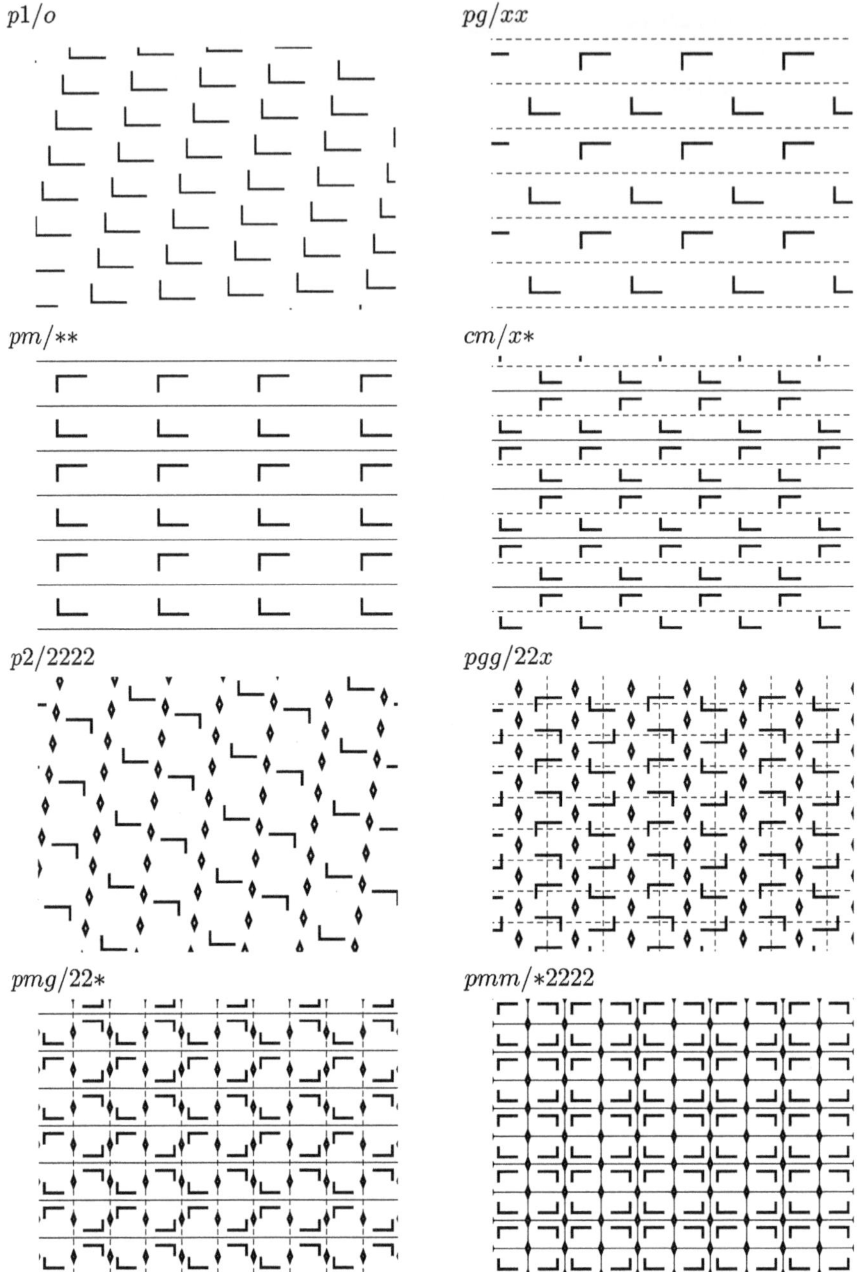

Fig. D.6 The seventeen wallpapers groups. Symbols are given in *Crystallographic/Orbifold* notation

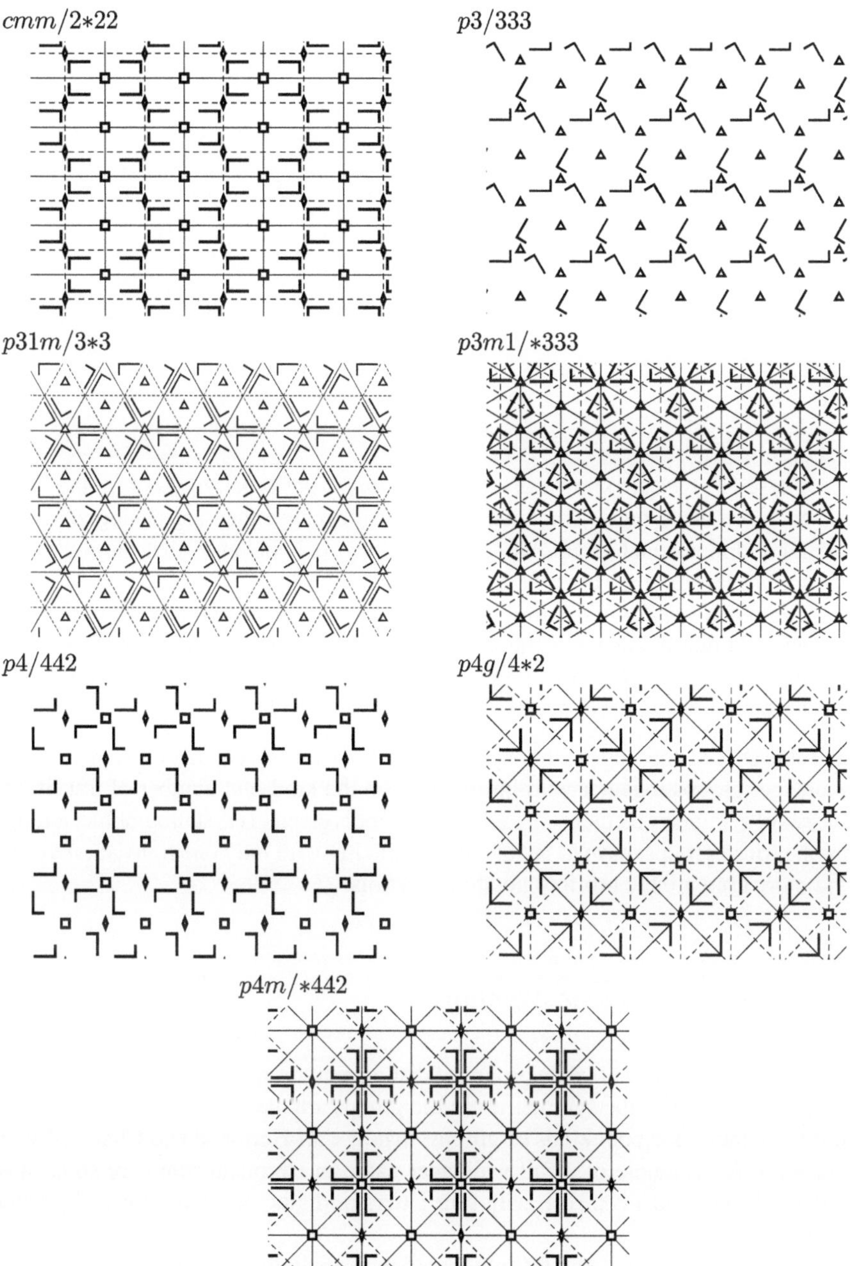

Fig. D.7 The seventeen wallpapers groups. Symbols are given in *Crystallographic/Orbifold* notation

$p6/632$ $p6m/*632$

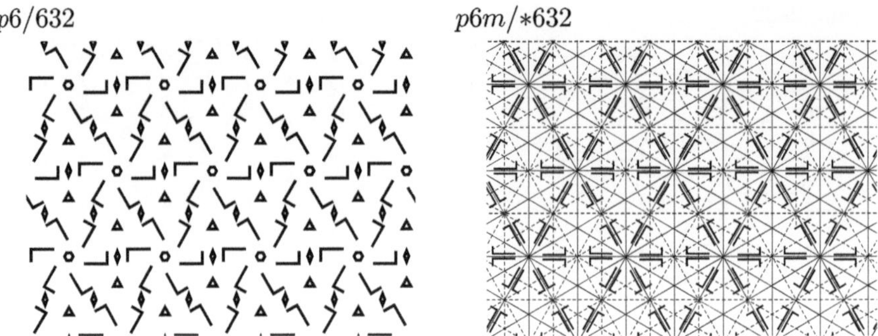

Fig. D.8 The seventeen wallpapers groups. Symbols are given in *Crystallographic/Orbifold* notation

These 17 types are known as *wallpaper groups, periodic groups* or (*plane*) *crystallographic groups*. The latter name arose because they are analogous to the three-dimensional groups of crystallography. It is believed that the *wallpaper groups* terminology arose as a result of an early paper [8]. Associated with each group is a *symbol* first introduced by crystallographers and now so universally adopted that is known as the *international symbol* see [7, 9], a brief introduction to these symbols will be given in Sect. D.4.1 and then another very useful terminology, the *orbifold notation*, is introduced in D.4.2.

In Table D.2, at the end of this appendix, we list the seven frieze groups and 17 wallpaper groups together with information on the kind ond number of transitivity classes of symmetry elements in each. The symbols given in crystallographic notation for the frieze groups, have to be intended implicitly with only one translation vector, and more used are the orbifold notation symbols.

D.4.1 Crystallographic Notation

Crystallography has 230 space groups to distinguish, far more than the 17 wallpaper groups, but many of the symmetries in the groups are the same. Thus we can use a similar notation for both kinds of groups, that of Carl Hermann and Charles-Victor Mauguin. An example of a full wallpaper name in Hermann-Mauguin style (also called *IUC notation*) is p31m, with four letters or digits; more usual is a shortened name like *cmm* or *pg*.

For wallpaper groups the full notation begins with either *p* or *c*, for a *primitive cell* or a *face-centered cell*; these are explained below. This is followed by a digit, n, indicating the highest order of rotational symmetry: 1-fold (none), 2-fold, 3-fold, 4-fold, or 6-fold. The next two symbols indicate symmetries relative to one translation axis of the pattern, referred to as the "main" one; if there is a mirror perpendicular to a translation axis we choose that axis as the main one (or if there are two, one of

Table D.2 Frieze and wallpaper groups

Symbol (1)		f or w (2)	Number of transitivity classes (3)					
					Reflections	rotations			
Crystallographic notation	Orbifold notation		Glide-reflections		Reflections	2	3	4	6
$p1$	$\infty\infty$	f	0		0	0	0	0	0
$p11g$	∞x	f	1		0	0	0	0	0
$p11m$	$\infty*$	f	0		1	0	0	0	0
$p1m1$	$*\infty\infty$	f	0		2	0	0	0	0
$p2$	22∞	f	0		0	2	0	0	0
$p2mg$	$2*\infty$	f	1		1	1	0	0	0
$p2mm$	$*22\infty$	f	0		3	2	0	0	0
$p1$	o	w	0		0	0	0	0	0
pg	xx	w	2		0	0	0	0	0
pm	$**$	w	0		2	0	0	0	0
cm	$x*$	w	1		1	0	0	0	0
$p2$	2222	w	0		0	4	0	0	0
pgg	$22x$	w	2		0	2	0	0	0
pmg	$22*$	w	1		1	2	0	0	0
pmm	$*2222$	w	0		4	4	0	0	0
cmm	$2*22$	w	2		2	2	0	0	0
$p3$	333	w	0		0	0	3	0	0
$p3lm$	$3*3$	w	1		1	0	2	0	0
$p3ml$	$*333$	w	1		1	0	1	0	0
$p4$	442	w	0		0	1	0	2	0
$p4g$	$4*2$	w	2		1	1	0	1	0
$p4m$	$*442$	w	1		3	1	0	2	0
$p6$	632	w	0		0	1	1	0	1
$p6m$	$*632$	w	2		2	1	1	0	1

(1) symbols in Crystallographic notation and orbifold notation; (2) indicates whether it is a frize or a wallpaper group; (3) give the number of transitivity classes of elements of the each kind

them). The symbols are either m, g, or 1, for mirror, glide reflection, or none. The axis of the mirror or glide reflection is perpendicular to the main axis for the first letter, and either parallel or tilted $180°/n$ (when $n > 2$) for the second letter. Many groups include other symmetries implied by the given ones. The short notation drops digits or an m that can be deduced, so long as that leaves no confusion with another group.

A *primitive cell* is a minimal region repeated by lattice translations. All but two wallpaper symmetry groups are described with respect to primitive cell axes, a coordinate basis using the translation vectors of the lattice. In the remaining two cases symmetry description is with respect to *centered cells* that are larger than the primitive cell, and hence have internal repetition; the directions of their sides is different from those of the translation vectors spanning a primitive cell. Hermann-Mauguin notation for crystal space groups uses additional cell types.

Examples

- *p2* (*p211*): Primitive cell, 2-fold rotation symmetry, no mirrors or glide reflections.
- *p4g* (*p4gm*): Primitive cell, 4-fold rotation, glide reflection perpendicular to main axis, mirror axis at 45o.
- *cmm* (*c2mm*): Centered cell, 2-fold rotation, mirror axes both perpendicular and parallel to main axis.
- *p31m* (*p31m*): Primitive cell, 3-fold rotation, mirror axis at 60°.

D.4.2 Orbifold Notation

Orbifold notation for wallpaper groups, introduced by John Horton Conway [10, 11] is based not on crystallography, but on topology. We first fold the infinite periodic tiling of the plane into its essence, an orbifold, then describe that with a few symbols:

- A digit, n, indicates a center of n-fold rotation corresponding to a cone point on the orbifold. By the crystallographic restriction theorem, n must be 2, 3, 4, or 6.
- An asterisk, $*$, indicates a mirror symmetry corresponding to a boundary of the orbifold. It interacts with the digits as follows:

  1. Digits before $*$ denote centers of pure rotation (cyclic group).
  2. Digits after $*$ denote centers of rotation with mirrors through them, corresponding to "corners" on the boundary of the orbifold (dihedral group).

- A cross, x, occurs when a glide reflection is present and indicates a cross-cap on the orbifold. Sometimes pure mirrors combine with lattice translation to produce glides, we do not take into account for these glides in the notation.
- the symbol ∞ indicates infinite rotational symmetry around a line; it can only occur for group of symmetries of Euclidean 3-space. By abuse of language, we might say that such a group is a subgroup of symmetries of the Euclidean plane with only one independent translation. The frieze groups occur in this way.
- The "no symmetry" symbol, o, stands alone, and indicates we have only lattice translations with no other symmetry. The orbifold with this symbol is a torus; in general the symbol, o denotes a handle on the orbifold.

Consider the group denoted in crystallographic notation by *cmm*, you can see it in Fig. D.7; in Conway's notation, this will be $2 * 22$. The 2 before the $*$ says we have a 2-fold rotation center with no mirror through it. The $*$ itself says we have a mirror. The first 2 after the $*$ says we have a 2-fold rotation center on a mirror. The final 2 says we have an independent second 2-fold rotation center on a mirror, one that is not a duplicate of the first one under symmetries.

The group denoted by *pgg* will be $22x$. We have two pure 2-fold rotation centers, and a glide reflection axis. Contrast this with *pmg*, Conway $22*$, where crystallographic notation mentions a glide, but one that is implicit in the other symmetries of the orbifold.

D.4.3 Why There Are Exactly Seventeen Groups

The orbifold notation gives us a tool to determine the number of wallpaper groups via a simple enumeration. An orbifold can be viewed as a polygon with face, edges, and vertices, which can be unfolded to form a possibly infinite set of polygons which tile either the sphere, the plane or the hyperbolic plane. When it tiles the plane it will give a wallpaper group and when it tiles the sphere or hyperbolic plane it gives either a spherical symmetry group or Hyperbolic symmetry group. The type of space the polygons tile can be found by calculating the Euler characteristic, $\chi = V - E + F$, where V is the number of corners (vertices), E is the number of edges and F is the number of faces. If the Euler characteristic is positive then the orbifold has a elliptic (spherical) structure; if it is zero then it has a parabolic structure, i.e. a wallpaper group; and if it is negative it will have a hyperbolic structure. When the full set of possible orbifolds is enumerated it is found that only 17 have Euler characteristic 0.

When an orbifold replicates by symmetry to fill the plane, its features create a structure of vertices, edges, and polygon faces, which must be consistent with the Euler characteristic. Reversing the process, we can assign numbers to the features of the orbifold, but fractions, rather than whole numbers. Because the orbifold itself is a quotient of the full surface by the symmetry group, the orbifold Euler characteristic is a quotient of the surface Euler characteristic by the order of the symmetry group.

The orbifold Euler characteristic is 2 minus the sum of the feature values, assigned as follows:

- A digit n before a $*$ counts as $\frac{n-1}{n}$.
- A digit n after a $*$ counts as $\frac{n-1}{2n}$.
- Both $*$ and x count as 1.
- The "no symmetry" o counts as 2.

For a wallpaper group, the sum for the characteristic must be zero; thus the feature sum must be 2.

References

1. B. Grunbaum, G.C. Shephard, *Tilings and Patterns* (W.H. Freeman and Company, New York, 1987)
2. F.H. Bool, J.R. Kist, F. Wierda, *M.C. Escher: His Life and Complete Graphic Work* (Harry N. Abrams, Inc., New York, 1992)
3. G. Pólya, Über die analogie der kristallsymmetrie in der ebene. Z. Kristallogr. **60**, 278–282 (1924)
4. D. Schattschneider, *Visions of Symmetry: Notebooks, Periodic Drawings, and Related Works of M.C. Escher* (W.H. Freeman and Company, New York, 1990)
5. G.E. Martin, *Transformation Geometry: An Introduction to Symmetry* (Springer, New York, 1982)
6. B.I. Rose, R.D. Stafford, An elementary course in mathematical symmetry. Am. Math. Monthly **88**, 59–64 (1981)

7. D. Schattschneider, The plane symmetry groups: their recognition and notation. Am. Math. Monthly **85**, 439–450 (1978)
8. M. Buerger, J. Lukesh, Wallpaper and atoms: how a study of nature's crystal aids scientist and artist. Technology Review, pp. 338–342 (1937)
9. H.S.M. Coxeter, W.O.J. Moser, *Generators and Relations for Discrete Groups*, 3rd edn. (Springer, Berlin, 1972)
10. J.H. Conway, The orbifold notation for surface groups, in *Groups, Combinatorics and Geometry*, ed. by M.W. Liebeck, J. Sax (Cambridge University Press, Cambridge, 1992), pp. 438–447
11. J.H. Conway, *The Symmetries of Things* (Peters Ltd, Wellesley, 2008)

About the Author

Guglielmo Paoletti is a post-doc researcher at the Laboratoire d'Infor- matique de Paris 6 (LIP6) at Université Pierre et Marie Curie since January 2013. He was born in 1982 in Varallo, Italy. He did his studies at University of Milan where he graduated in Physics in 2007. He moved to Pisa where he got his doctorate in Physics at University of Pisa in 2012. He was student of Prof. Sergio Caracciolo and collaborated with Andrea Sportiello. After the PhD he has been a post-doc, sustained by *Fondazione Angelo della Riccia*, at the Laboratory of Theoretical Physics and Statistical Models (LPTMS) in Orsay-Paris in 2012. During that period he continued his research on the Abelian Sandpile Model and started a collaboration with Alberto Rosso, focused on the case of Fixed Energy Sandpile. In 2012 he has been vis- iting Cornell University (Ithaca, USA) and Tata Insitute of Fundamental Research (Mumbai, India). He is interested in Statistical Mechanics and Critical Phenomena in relation with combinatorics and graph theory. He published several peer-reviewed papers on international journals

G. Paoletti, *Deterministic Abelian Sandpile Models and Patterns*,
Springer Theses, DOI: 10.1007/978-3-319-01204-9,
© Springer International Publishing Switzerland 2014

If you have any concerns about our products,
you can contact us on
ProductSafety@springernature.com

In case Publisher is established outside the EU,
the EU authorized representative is:
Springer Nature Customer Service Center GmbH
Europaplatz 3, 69115 Heidelberg, Germany

Printed by Libri Plureos GmbH
in Hamburg, Germany